D1258270

INSIDE THE HUMAN BODY

3 1526 03327586 5

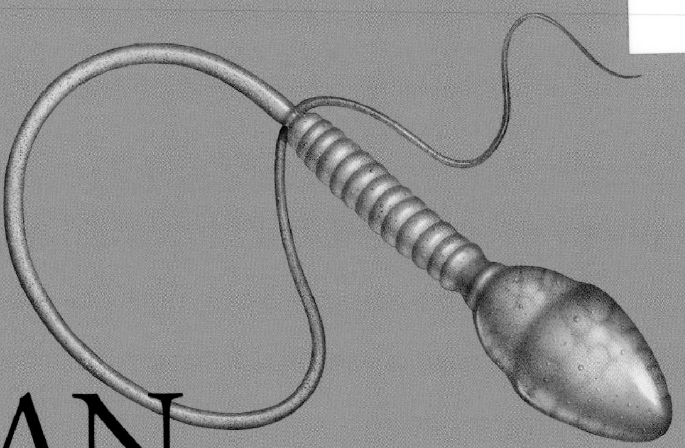

HUMAN REPRODUCTION AND DEVELOPMENT

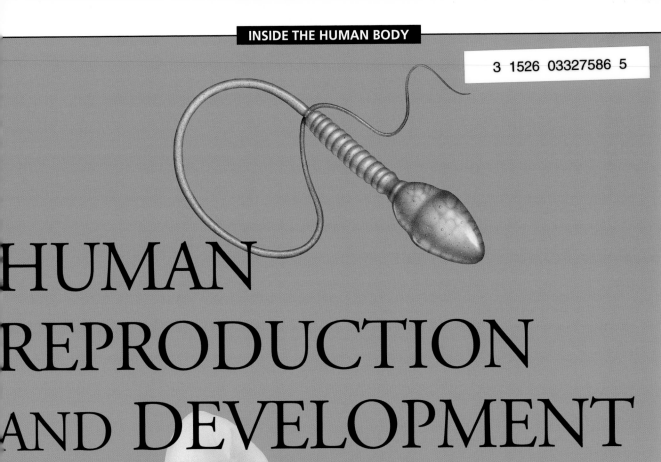

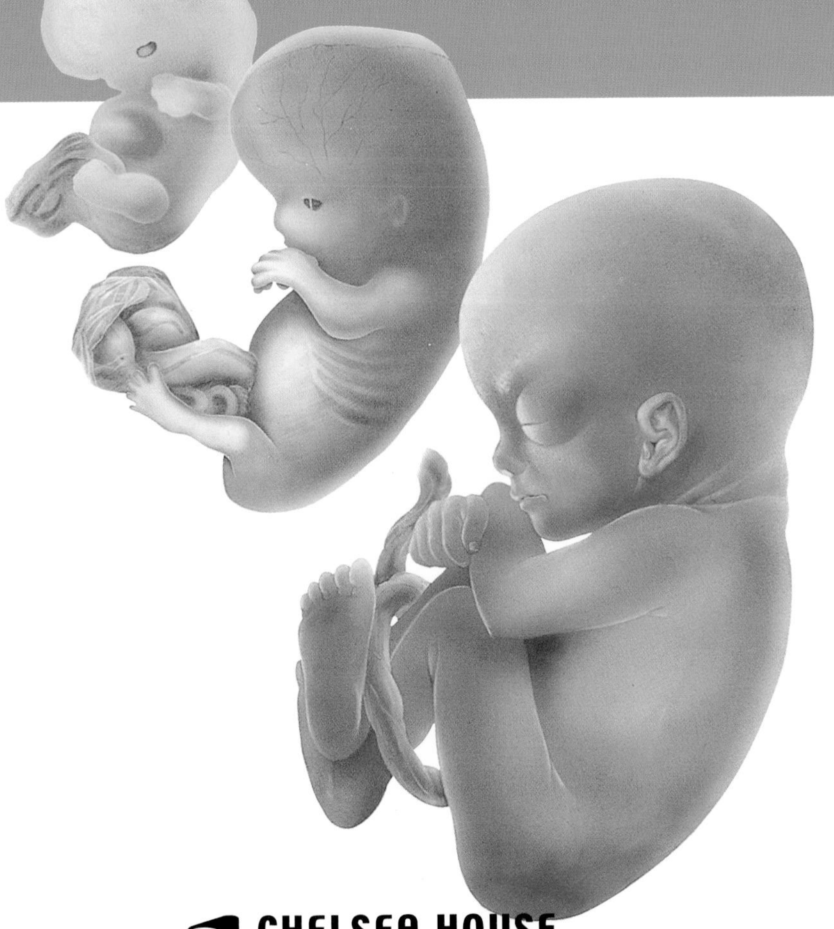

CHELSEA HOUSE PUBLISHERS
A Haights Cross Communications Company ®
Philadelphia

HARFORD COUNTY
PUBLIC LIBRARY
100 E. Pennsylvania Avenue
Bel Air, MD 21014

First hardcover library edition published
in the United States of America in
2006 by Chelsea House Publishers,
a subsidiary of Haights Cross Communications.
All rights reserved.

A Haights Cross Communications Company ®

www.chelseahouse.com

Library of Congress Cataloging-in-Publication
applied for.
ISBN 0-7910-9015-9

Project and realization
Parramón, Inc.

Texts
Adolfo Cassan

Translator
Patrick Clark

Graphic Design and Typesetting
Toni Inglés Studio

Illustrations
Marcel Socías Studio

First edition - September 2004

Printed in Spain
© Parramón Ediciones, S.A. – 2005
Ronda de Sant Pere, 5, 4ª planta
08010 Barcelona (España)
Norma Editorial Group

www.parramon.com

The whole or partial reproduction of this work by any means or procedure, including printing, photocopying, microfilm, digitalization, or any other system, without the express written permission of the publisher, is prohibited.

TABLE OF CONTENTS

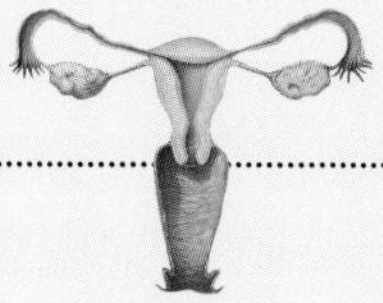

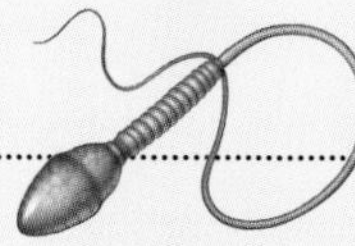

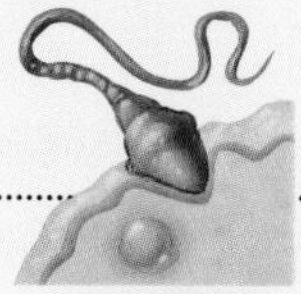

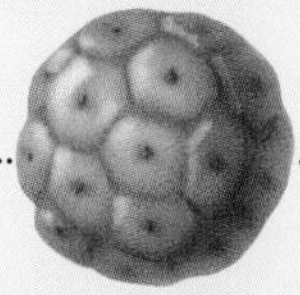

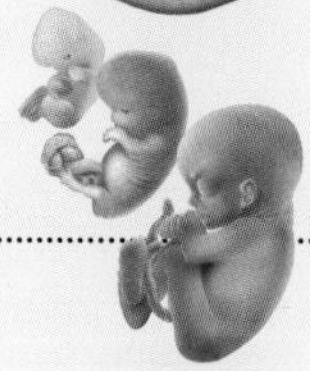

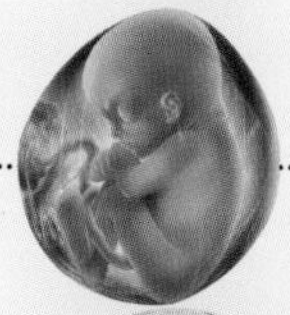

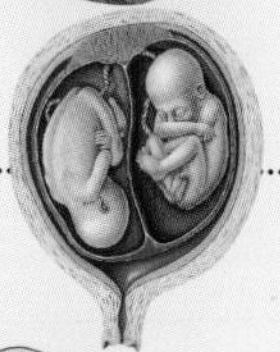

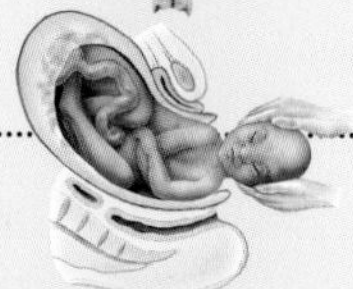

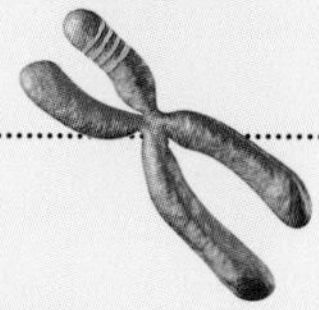

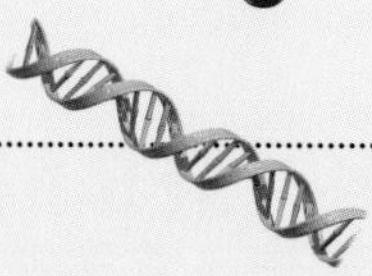

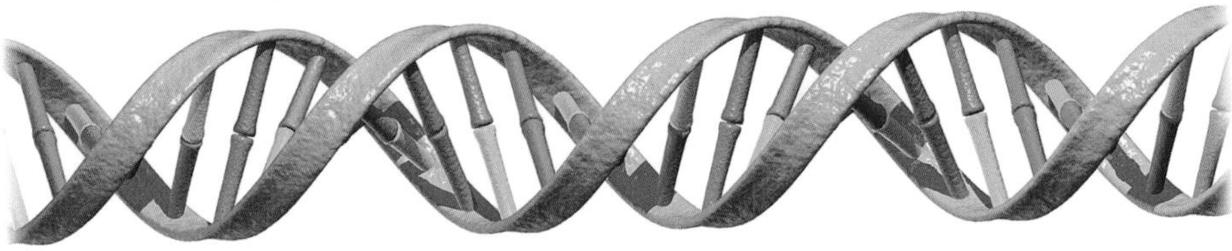

HOW WE ARE BORN

This book aims to give young readers some basic information about human reproduction, the amazing and complex process that leads to the birth of new human beings.

After a brief introduction that explains most of the general topics related to reproduction, this book deals with the most important aspects of the formation and development of a new human being, from its origin and growth inside the mother's womb until the moment of birth. Each section has a large illustration with some brief explanations about the most relevant points. At the end of the book, a few interesting facts are explained.

Our goal was to create a book that is practical, instructive, challenging, and enjoyable. We hope our readers will consider this goal accomplished.

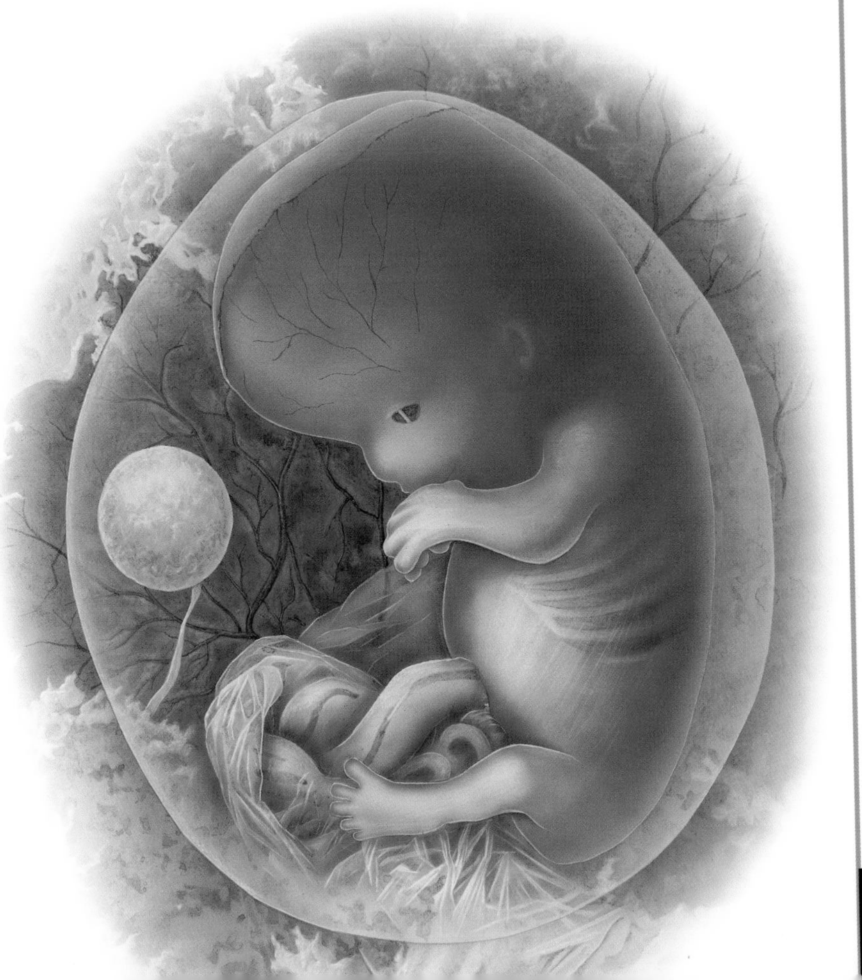

HUMAN REPRODUCTION

A NATURAL PROCESS

The meeting of two tiny germ cells called the egg cell and the sperm cell marks the start of a series of amazing events. This meeting results in the formation of a single cell, known as the zygote, that, over the course of nine months or so inside the mother's body, will become a baby. This is a very natural and ordinary event. Indeed, there is nothing more natural than the development of a new human being inside its mother's body. This is life.

We live in a world with more than 6 billion people, and all of us, along with our ancestors, started in this way. This is how our species reproduces, and our bodies are specially adapted for this purpose.

HUMAN BEINGS, A SPECIES WITH TWO GENDERS

Reproduction is the process by which all living things create life similar to themselves. But there are many differer kinds of reproduction. Bacteria, for example, multiply by means of simple cellular division. Many animals also reproduce by layin eggs from which their young wil be born. Others conceive their offspring inside the mother' body, where the developi embryo will be sheltere until the moment of birt Among these viviparous ("live birth") animals a human beings.

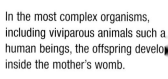

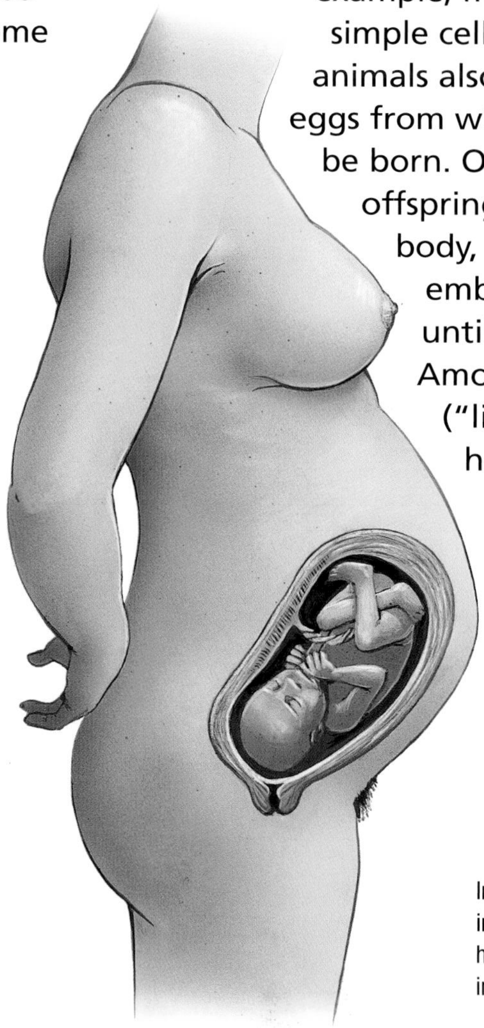

In the most complex organisms, including viviparous animals such a human beings, the offspring develo inside the mother's womb.

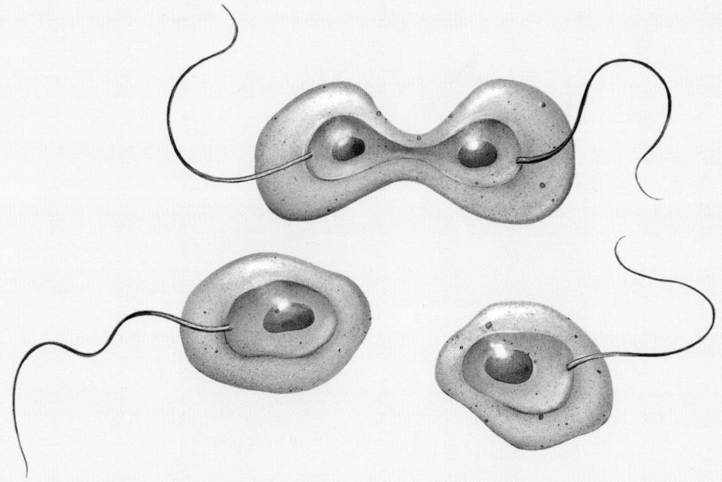

The simplest organisms have a very basic form of reproduction. Bacteria reproduce by means of cellular division, which creates two bacteria exactly the same as the first.

allow human reproduction to occur, ere must be two genders. In scientific rms, this is called "sexual dimorphism." other words, reproduction requires the istence of individuals of the male and male sex. The differences between the odies of men and women allow them to ay distinct and complementary roles in e reproductive process. These fferences are most notable in the nitals, because the reproductive system each sex is designed to produce a fferent germ cell. Both germ cells meet the act of copulation.

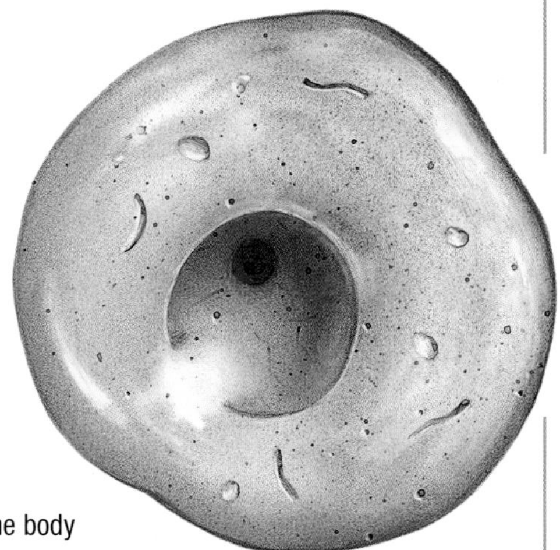

cells of the human y contain 46 omosomes—23 s—in their nuclei. se chromosomes vide the genetic rmation needed for the body rm and function.

GERM CELLS

The cells of the human body contain 46 chromosomes in their nuclei. Chromosomes are tiny particles that carry the genetic information needed for the formation and functioning of cells and the body as a whole. There are 23 pairs of chromosomes, which correspond to each other in pairs of similar size and shape, except for the pair that holds the sex chromosomes. Sex chromosomes are the same in females (XX) and different in males (XY).

Male germ cells, produced in the testicles, are called spermatozoa. During puberty (between the ages of 10 and 14), and throughout the entire adult life of men, the testicles are constantly producing millions of sperm cells. Female germ cells are called egg cells or oocytes. The ovaries of a newborn girl contain hundreds of thousands of oocytes. Beginning in puberty and continuing in cycles during the adult life of a woman, just a few hundred of these egg cells will mature. Of course, for a human to be conceived, one crucial step is necessary: A male germ cell and a female germ cell must meet and fuse

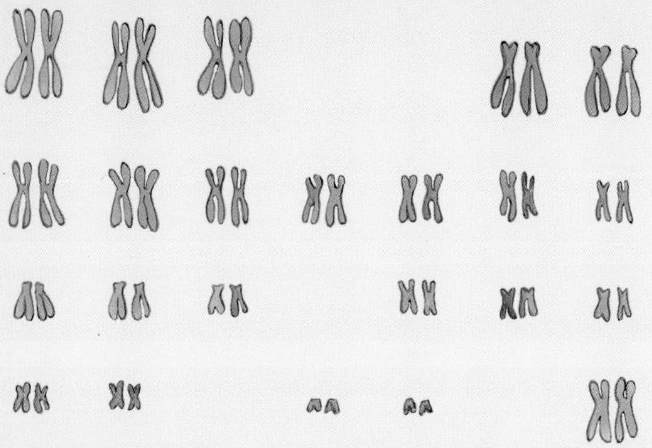

Human beings have 46 chromosomes that can be placed into 23 pairs according to their size and shape. Here, we can see the chromosome set of a woman, known as the female karyotype.

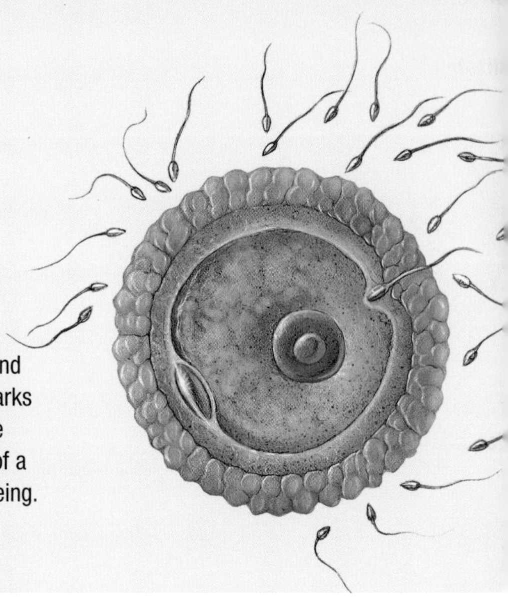

The meeting and fusion of a sperm cell and an egg cell marks the start of the development of a new human being.

together, each bringing its 23 chromosomes to create a new cell that is equipped with 46 chromosomes, for a complete set of chromosomes with the instructions needed for the formation of a new life.

THE FIRST STEP: FERTILIZATION

The genital organs of a man and a woman are specially adapted to carry out sexual intercourse, also known as coitus or copulation. In this union, the penis of the male and the vagina of the female fit together like pieces of a puzzle. During this process, a liquid made inside the body of the man is deposited inside the woman's body. Millions of spermatozoa float inside this liquid, which is known as semen. Immediately, the male germ cells enter the woman's reproductive organs in search of a female germ cell. It is possible that they will not find one, but it is also possible that they will. If this happens, then a union between the sperm cell and egg cell may occur. This event is called fertilization.

Fertilization results in the creation of a very special cell known as the zygote, which has two extraordinary characteristics. For one, it has a complete set of chromosomes, half coming from the sperm cell and half from the egg cell. It also has the potential to become a human being. If conditions are right, the zygote can grow into a complete human being, with its very own unique characteristics.

BRYONIC STAGE

lization

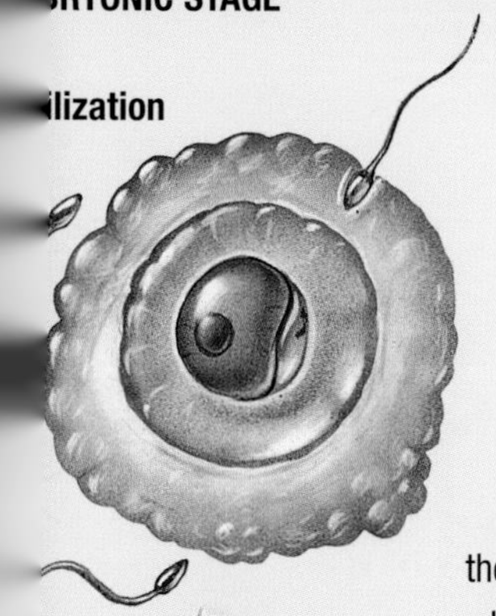

lantation

bryo

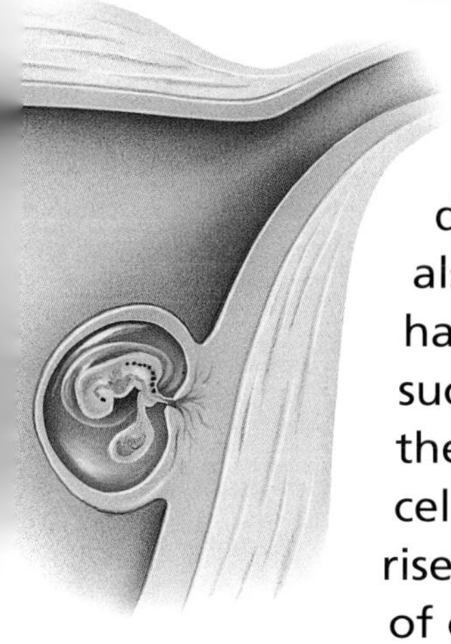

ETAL STAGE

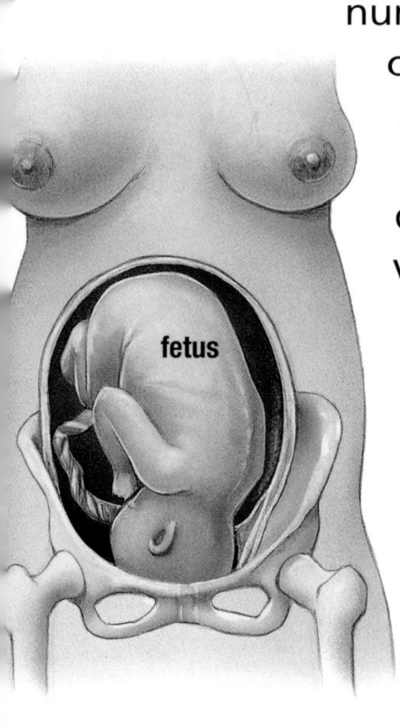

It takes nine months for the fetus to develop. During this time, the zygote develops inside the mother's uterus, passing through an embryonic stage, then a fetal stage, until it becomes mature enough to survive outside the mother's body.

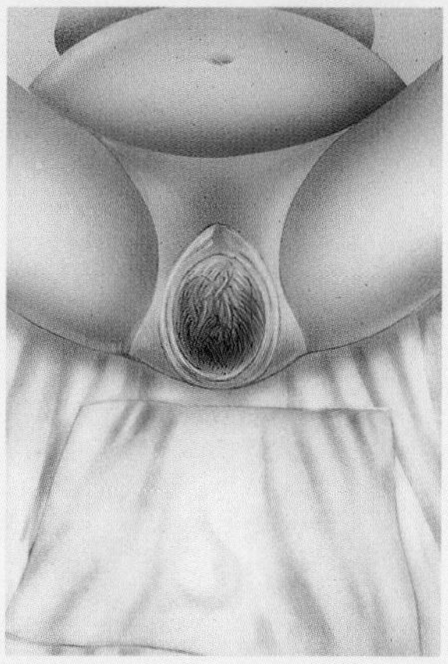

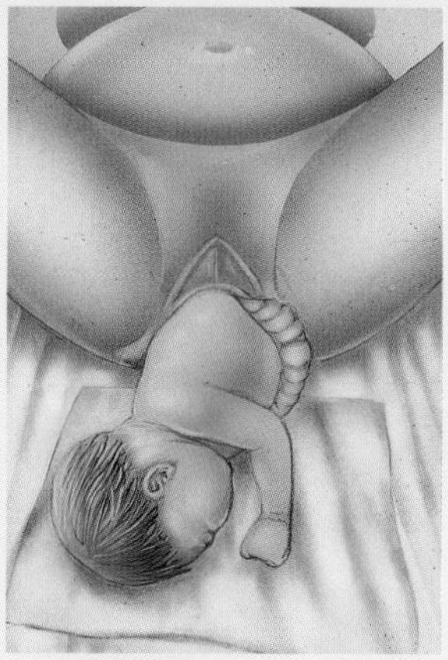

After nine months of waiting, the process of delivery begins, and the baby leaves its mother's womb to face the outside world.

THE DEVELOPMENT OF A NEW HUMAN BEING

Following the "instruction manual" of its genetic makeup, the zygote is divided into two identical daughter cells, which also divide. The same happens with their successors, and with the next generation of cells. A single cell gives rise to a large number of cells that have exactly the same number of chromosomes, but are no longer identical. These cells differentiate to form various structures and to carry out different functions. The zygote thus turns into an embryo, and then into a fetus that grows continually.

In a few months, millions of cells that come from the initial zygote form a complete human body. Some of the cells will make up the various layers and glands of the skin; others will become hair; some will form the liver, lungs, and heart; others will become red blood cells, or generate muscles, tendons, bones, and lungs.

Primitive cells will turn into a very complex and sophisticated human body—indeed, into something much more than that—a person that will be born about nine months after fertilization. The parents will be eager to know their baby, and ready to provide it with love and care. Even this parental attitude is encoded in our genetic information: For humans, as for so many members of the animal kingdom, our children, the future of our species, are the most important thing in the world.

THE GENITALS

Genitals are formed by a set of organs that take part in the process of procreation. The organs that make up the male genitals are designed to produce sperm cells that can fertilize an egg. The female genital organs produce eggs and also serve to nourish and shelter the developing fetus during pregnancy.

FEMALE GENITALS

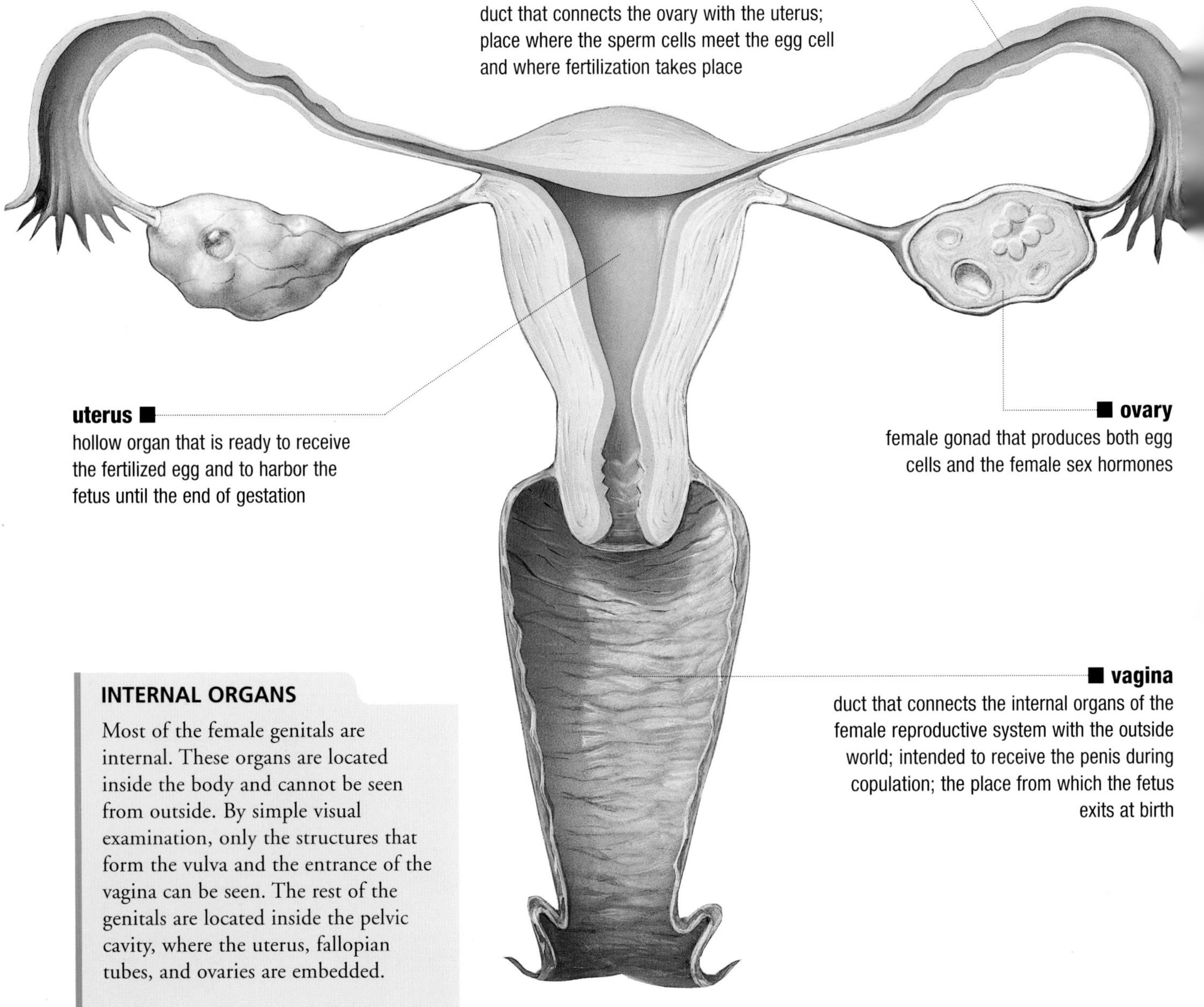
fallopian tube
duct that connects the ovary with the uterus; place where the sperm cells meet the egg cell and where fertilization takes place
uterus
hollow organ that is ready to receive the fertilized egg and to harbor the fetus until the end of gestation
ovary
female gonad that produces both egg cells and the female sex hormones
vagina
duct that connects the internal organs of the female reproductive system with the outside world; intended to receive the penis during copulation; the place from which the fetus exits at birth

INTERNAL ORGANS

Most of the female genitals are internal. These organs are located inside the body and cannot be seen from outside. By simple visual examination, only the structures that form the vulva and the entrance of the vagina can be seen. The rest of the genitals are located inside the pelvic cavity, where the uterus, fallopian tubes, and ovaries are embedded.

THE TESTICLES

The testicles, which make sperm cells, would not work well if they had to tolerate the warm temperature inside the body. For this reason, they are housed in the scrotum, a cutaneous (formed of skin) sack that hangs from the base of the penis, where they are exposed to a somewhat lower temperature that is ideal for their survival.

MALE GENITALS

s deferens e through which mature rm cells travel outside body

prostate gland that produces a secretion that is part of the semen, rich in nutrients for the sperm cells

seminal vesicle gland that makes part of the semen in which the sperm cells float

penis organ that deposits sperm cells in the vagina of a woman

testicle male gonad that produces sperm cells and makes male sex hormones

epididymis tube next to the testicle, inside of which sperm cells mature

urethra duct that crosses the penis and through which semen is forced out of the body

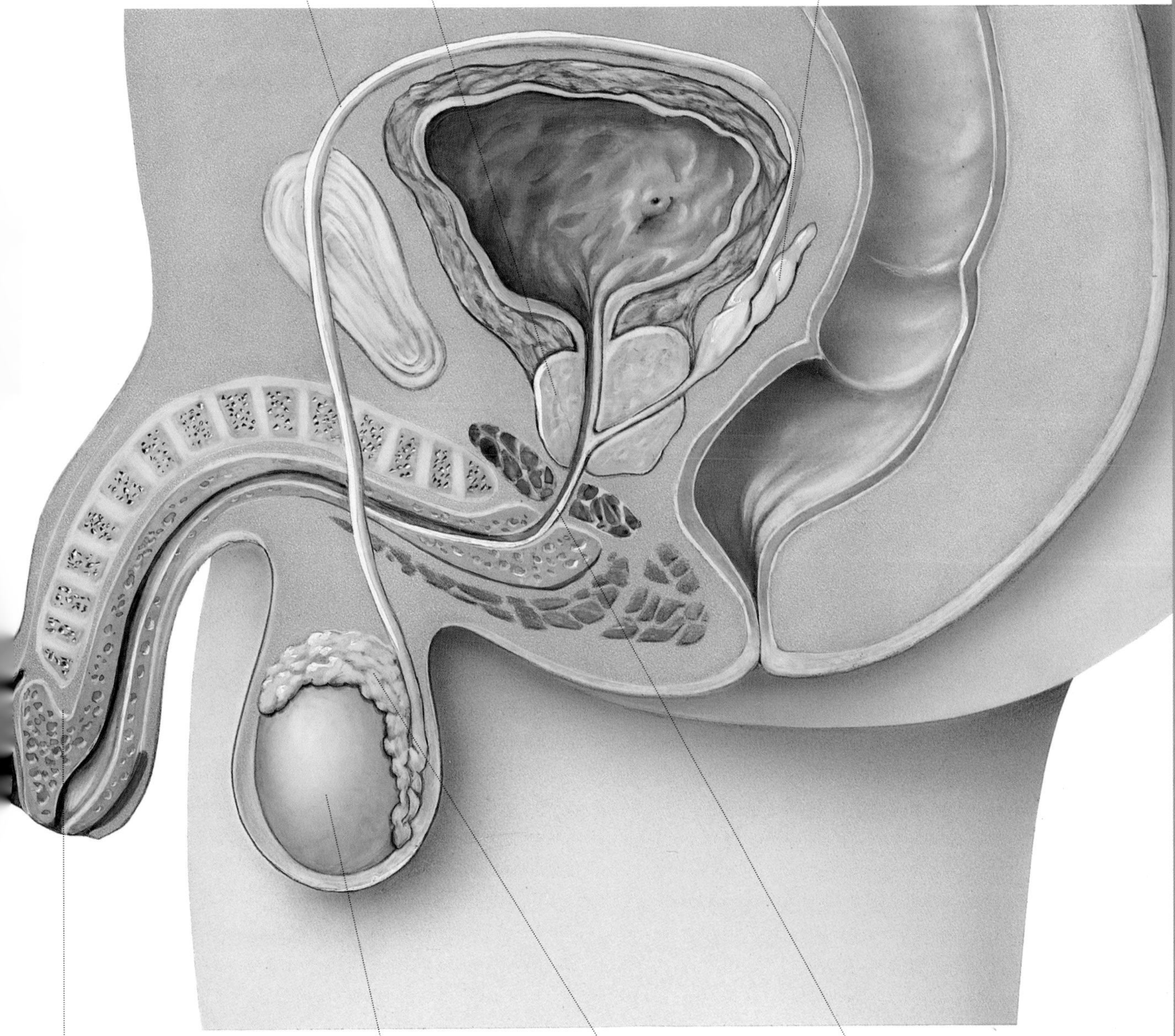

THE SPERM CELL AND THE EGG CELL

The sex cells, also known as gametes or germ cells, are different from the rest of the human body's cells, mainly because they have only 23 chromosomes, a peculiarity that makes their particular function possible. The union of the sperm cell—the male sex cell—and the egg cell—the female sex cell—will give rise to a complete cell, with 46 chromosomes, whose divisions will bring about the development of a new human being.

SPERM CELL
male sex cell

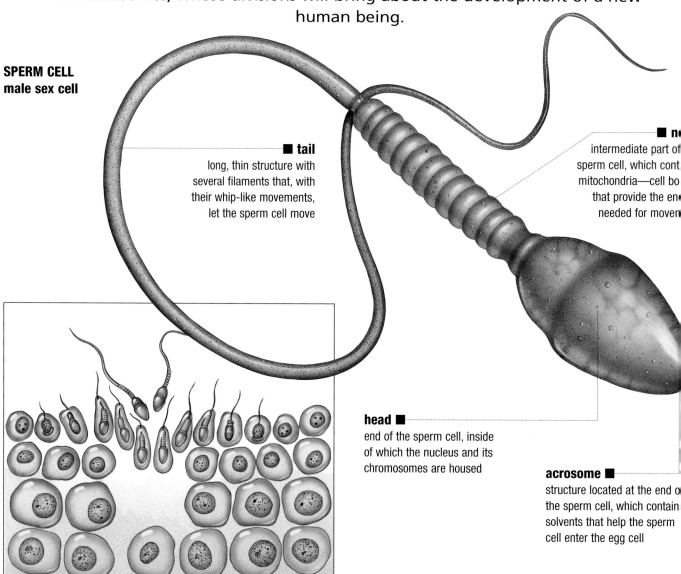

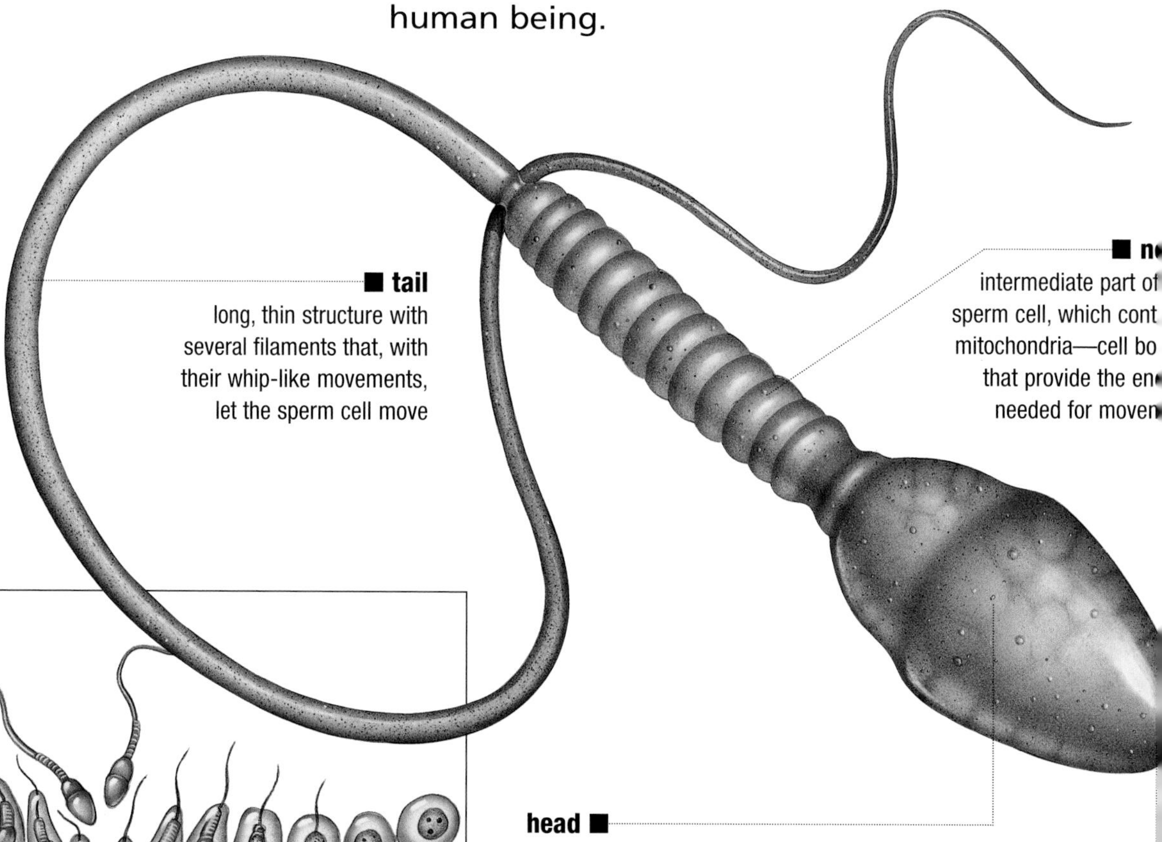
■ tail
long, thin structure with several filaments that, with their whip-like movements, let the sperm cell move
■ n
intermediate part of sperm cell, which cont mitochondria—cell bo that provide the en needed for moven
head ■
end of the sperm cell, inside of which the nucleus and its chromosomes are housed
acrosome ■
structure located at the end o the sperm cell, which contain solvents that help the sperm cell enter the egg cell

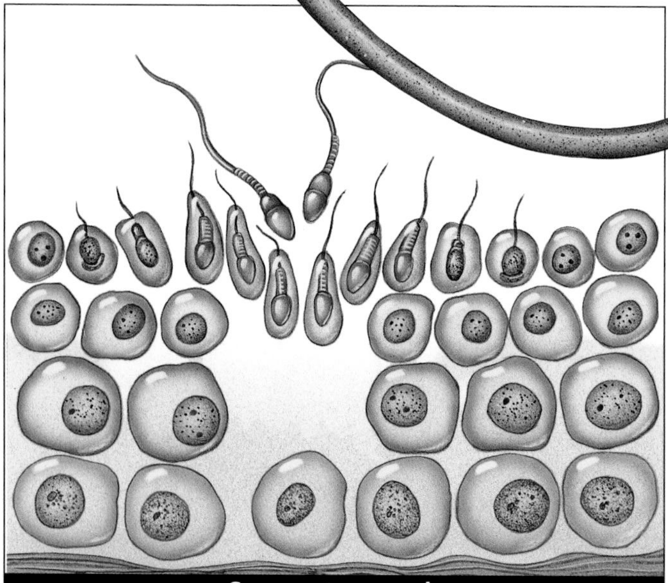

Spermatogenesis

Inside the testicles, frantic activity takes place, leading to the constant production of sperm cells. Thousands and thousands of tiny cells divide and mature to create the male sex cells. Thanks to this uninterrupted activity, called spermatogenesis, each cubic centimeter of semen may contain between 100 and 200 million sperm cells.

THE SPERM CELL

This is one of the smallest cells in the body: It measures only between 50 and 60 thousandths of a millimeter.

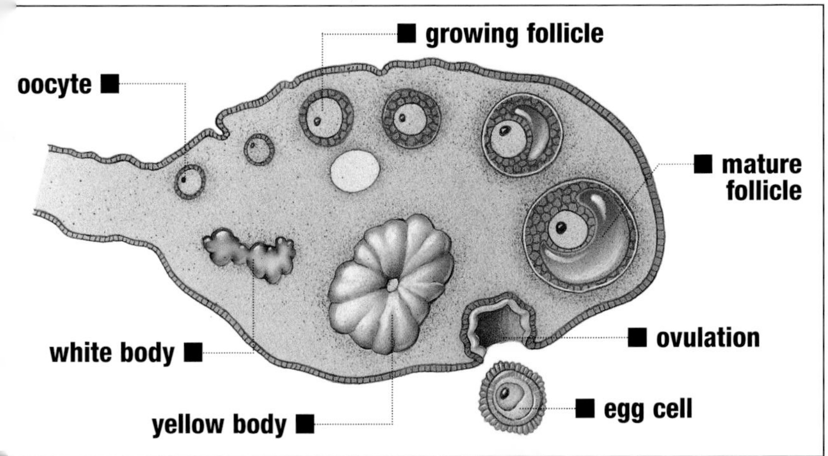
growing follicle
oocyte
mature follicle
white body
ovulation
egg cell
yellow body

Ovulation

ıe ovaries contain some 400,000 primary oocytes at the moment of birth, but only a tiny rtion of them will mature and change into egg cells over the course of life. Beginning with uberty, and during the entire reproductive life of a woman, a cycle of about one month in duration takes place. During this cycle, a follicle containing an oocyte grows. Toward the niddle of the cycle, it bursts and releases a mature egg cell that is ready to be fertilized.

THE EGG CELL

The mature egg cell is one of the largest cells in the body, enormous in comparison to the tiny sperm cell: It measures about one millimeter in diameter.

GG CELL
male sex cell

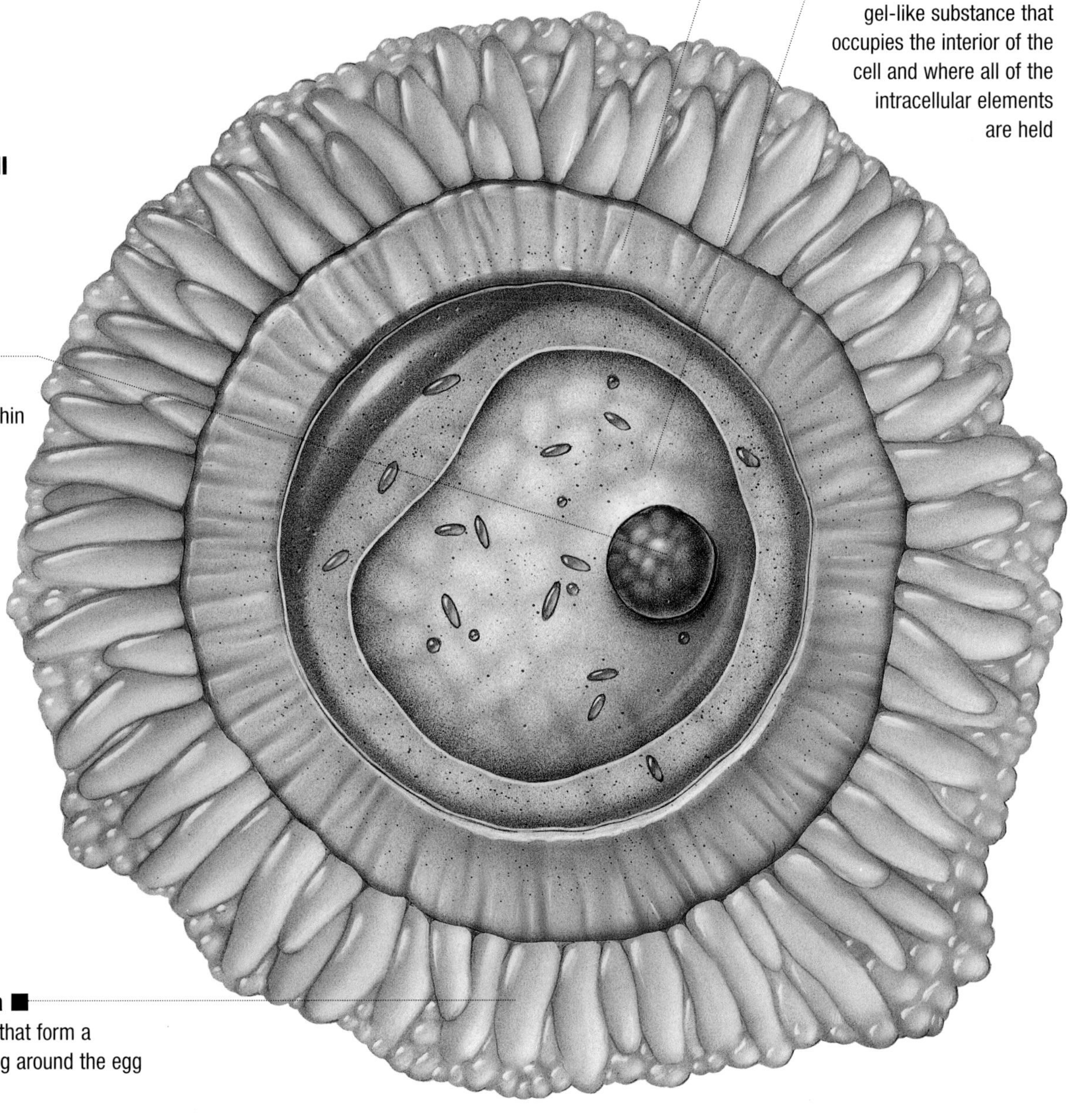
zona pellucida
thin but resistant membrane that surrounds the egg cell and makes up a physical protective barrier
cytoplasm
gel-like substance that occupies the interior of the cell and where all of the intracellular elements are held
ucleus
ellular structure, urrounded by a thin embrane, that ontains the hromosomes
corona radiata
set of small cells that form a protective covering around the egg cell

THE MEETING OF THE GAMETES

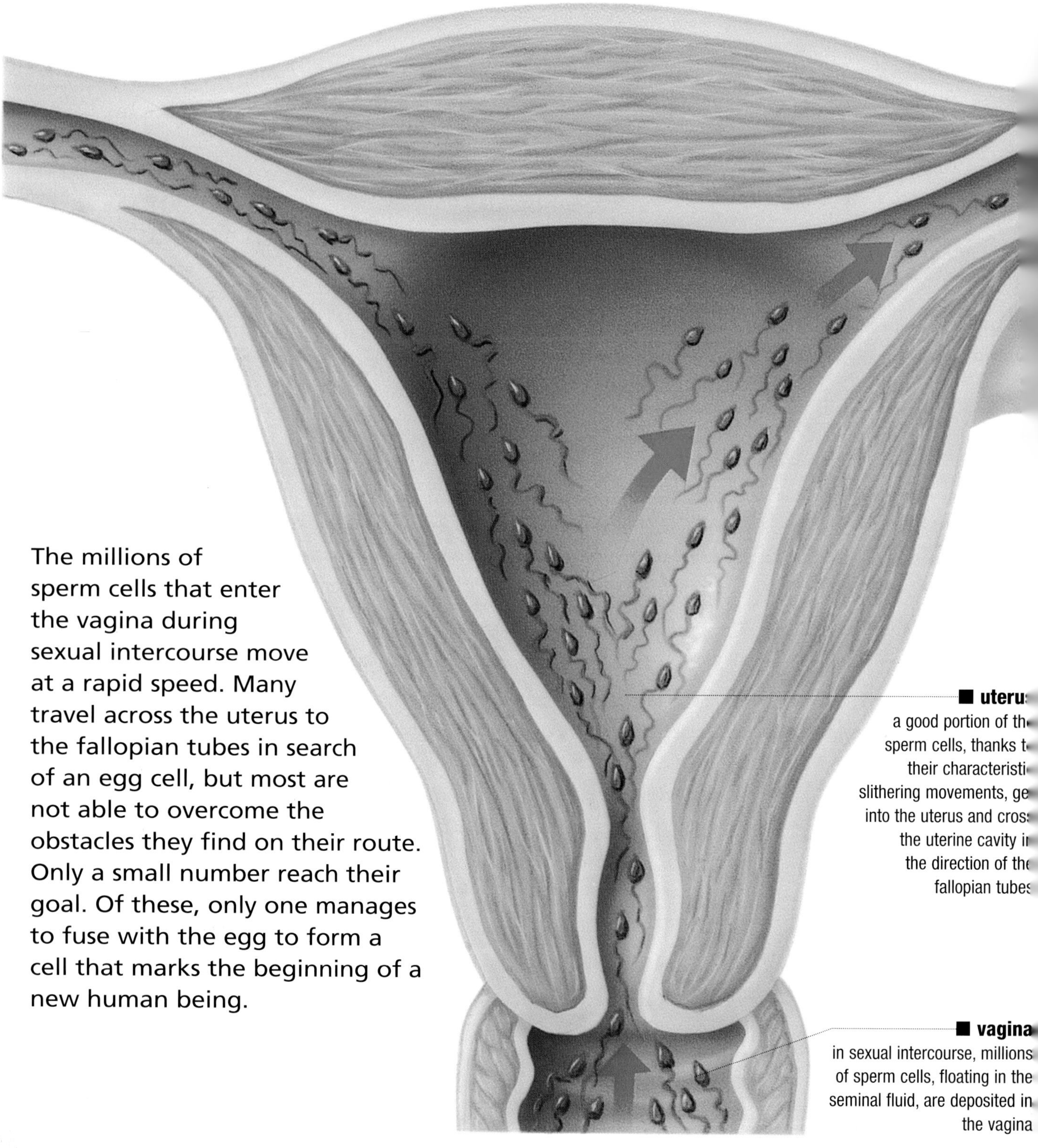

The millions of sperm cells that enter the vagina during sexual intercourse move at a rapid speed. Many travel across the uterus to the fallopian tubes in search of an egg cell, but most are not able to overcome the obstacles they find on their route. Only a small number reach their goal. Of these, only one manages to fuse with the egg to form a cell that marks the beginning of a new human being.

■ **uterus**
a good portion of the sperm cells, thanks to their characteristic slithering movements, get into the uterus and cross the uterine cavity in the direction of the fallopian tubes

■ **vagina**
in sexual intercourse, millions of sperm cells, floating in the seminal fluid, are deposited in the vagina

N *VITRO* FERTILIZATION

f a couple has trouble achieving pregnancy, today it is possible to try various medical ›rocedures to start a family. Among these techniques for reproduction is one called *n vitro* fertilization. When there are problems that prevent a union of the sperm cell ınd egg cell inside the mother's body, the germ cells are put together in a laboratory. \fter fertilization is achieved in a glass test tube, the embryo is then placed into the nother's uterus to continue its development.

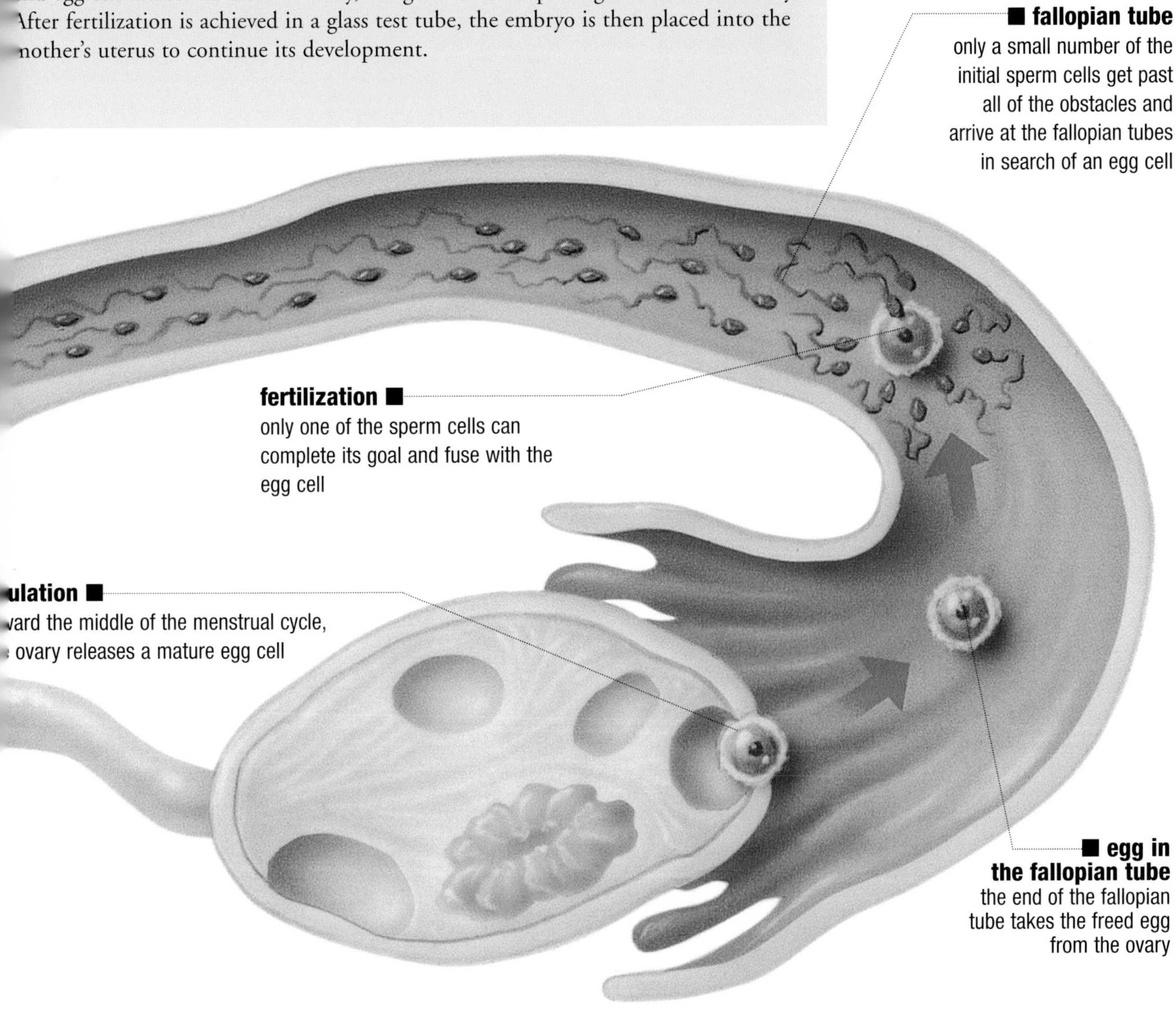
fallopian tube
only a small number of the initial sperm cells get past all of the obstacles and arrive at the fallopian tubes in search of an egg cell
fertilization
only one of the sperm cells can complete its goal and fuse with the egg cell
ulation
vard the middle of the menstrual cycle,
ovary releases a mature egg cell
egg in the fallopian tube
the end of the fallopian tube takes the freed egg from the ovary

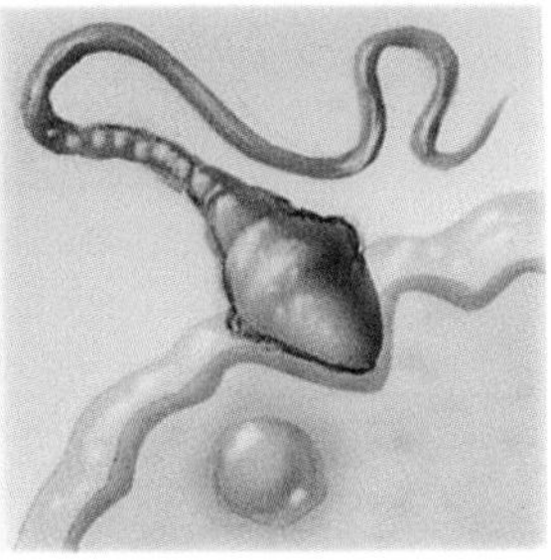

1. A sperm cell manages to cross the corona radiata that surrounds the egg cell, and is supported on its surface.

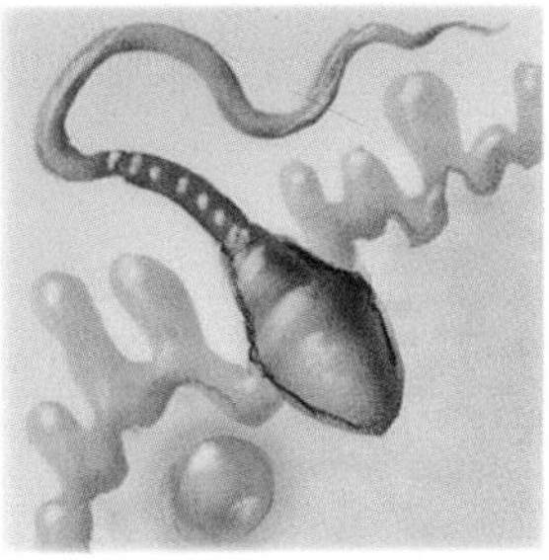

2. The sperm cell, due to the action of the solvent substances it carries in its head, perforates the membrane of the egg cell and begins to insert itself.

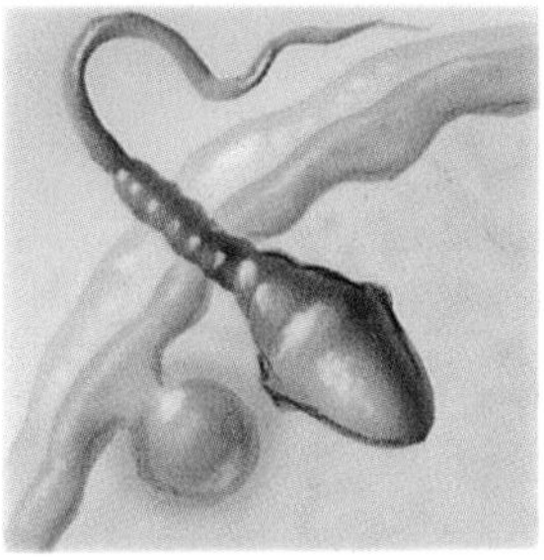

3. When the sperm cell manages to insert itself, the membrane of the egg cell begins to repair itself to stop other sperm cells from getting in.

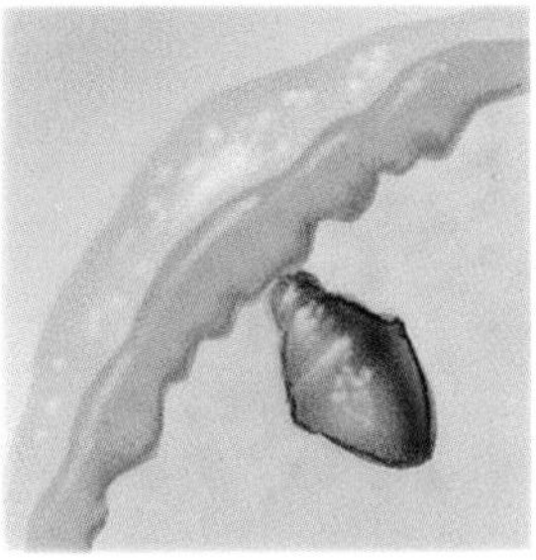

4. The tail of the sperm cell falls away and only the head, which contains the chromosomes that carry the genetic information, remains inside the egg cell.

The crucial moment, step by step

THE FIRST EMBRYONIC DIVISIONS

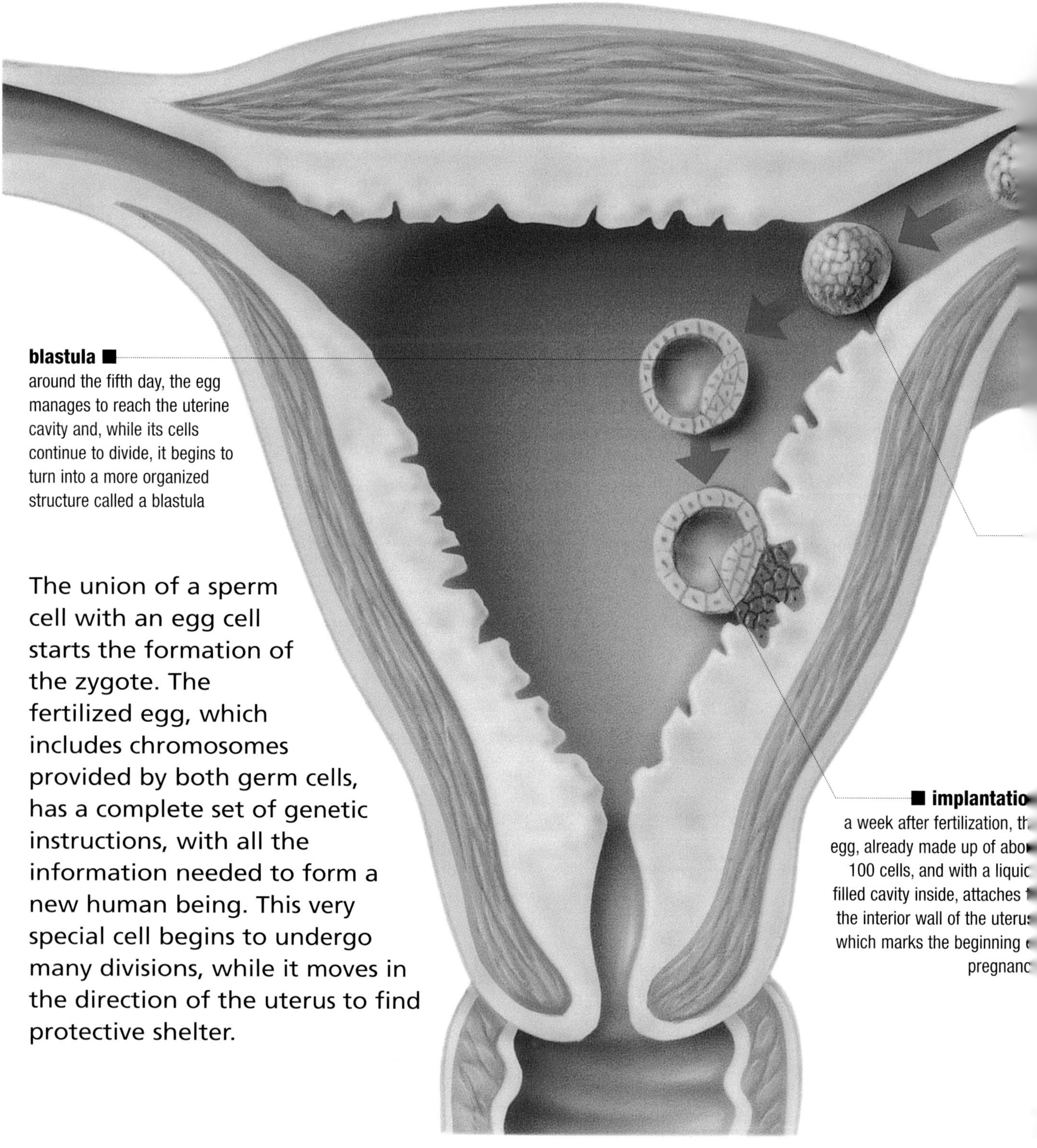

blastula ■
around the fifth day, the egg manages to reach the uterine cavity and, while its cells continue to divide, it begins to turn into a more organized structure called a blastula

The union of a sperm cell with an egg cell starts the formation of the zygote. The fertilized egg, which includes chromosomes provided by both germ cells, has a complete set of genetic instructions, with all the information needed to form a new human being. This very special cell begins to undergo many divisions, while it moves in the direction of the uterus to find protective shelter.

■ **implantatio**
a week after fertilization, th egg, already made up of abo 100 cells, and with a liquic filled cavity inside, attaches the interior wall of the uterus which marks the beginning pregnanc

N SEARCH OF A NEST

'he zygote has a supply of nutrients that allows it to live for only a few days. It needs to ind shelter quickly, before a week has passed. It is exactly at the end of seven days after ertilization that it makes contact with the internal wall of the uterus and carves itself a nest" in a process known as implantation.

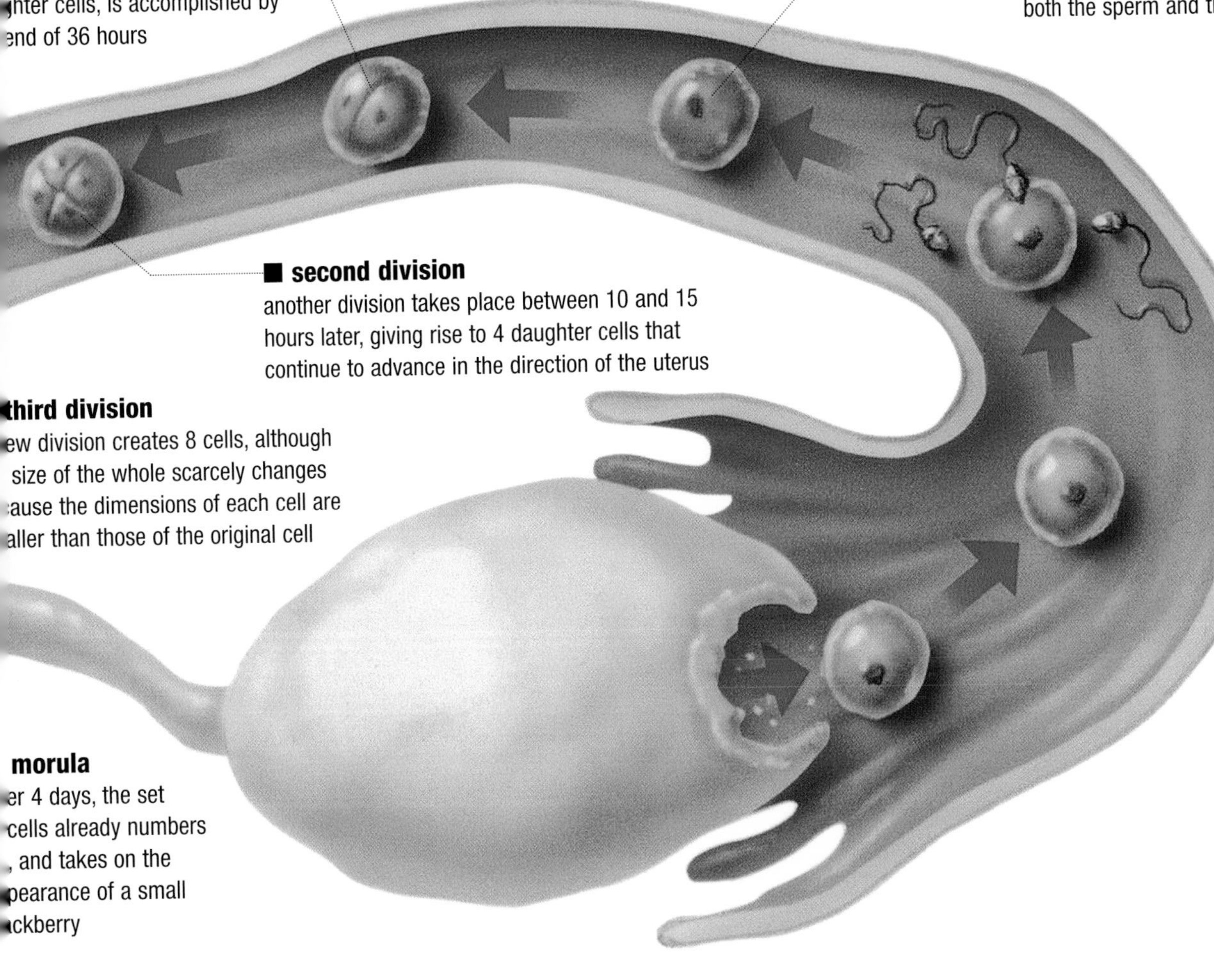

t division ■
irst division of the zygote, g rise to two identical ghter cells, is accomplished by end of 36 hours

■ zygote
fertilization of the egg results in the zygote, which has all of the elements of both the sperm and the egg

■ second division
another division takes place between 10 and 15 hours later, giving rise to 4 daughter cells that continue to advance in the direction of the uterus

third division
ew division creates 8 cells, although size of the whole scarcely changes ause the dimensions of each cell are aller than those of the original cell

morula
er 4 days, the set cells already numbers , and takes on the pearance of a small ckberry

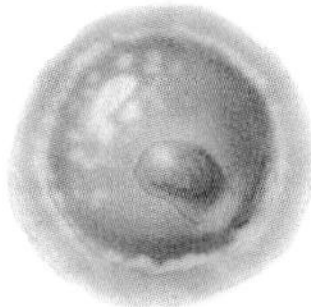

The egg cell is fertilized by a sperm cell

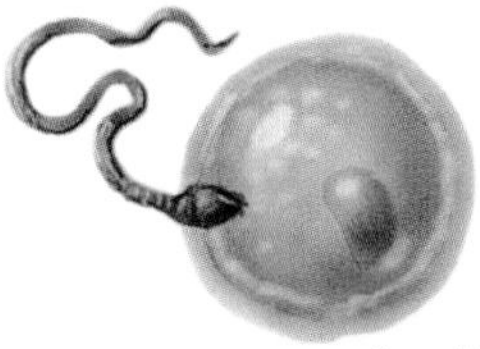

The fusing of the egg cell and the sperm cell creates a zygote

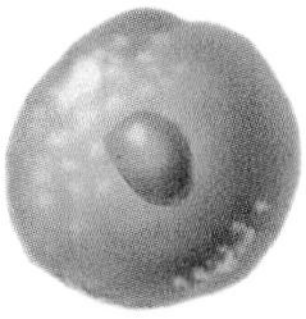

The zygote divides and two identical daughter cells form

Zygote divides into four cells

Zygote divides into eight cells

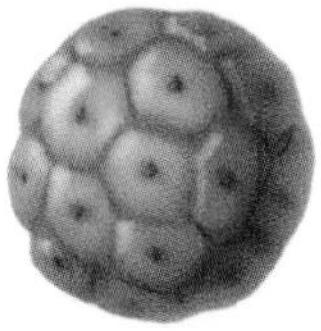

Morula, consisting of 32 cells

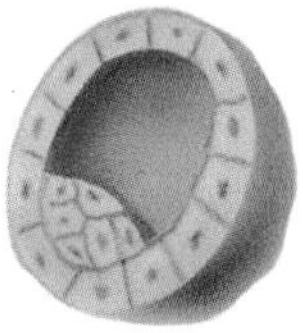

A progressive accumulation of fluid inside the egg creates a blastula

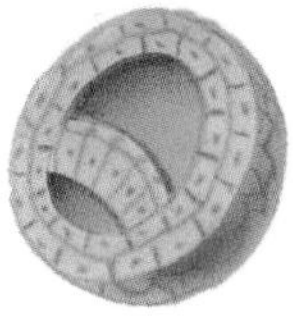

The blastula is implanted in the uterine wall

The first week, step by step

THE FORMATION OF A NEW HUMAN BEING

The first eight weeks of gestation are called the embryonic period, during which some spectacular changes take place: A simple group of cells, following instructions that are built into their genetic code, begin developing into different organic structures that form an embryo. This group of cells also begins to develop membranes that will protect the embryo and the placenta, an organ that supplies the embryo with oxygen and food as it develops.

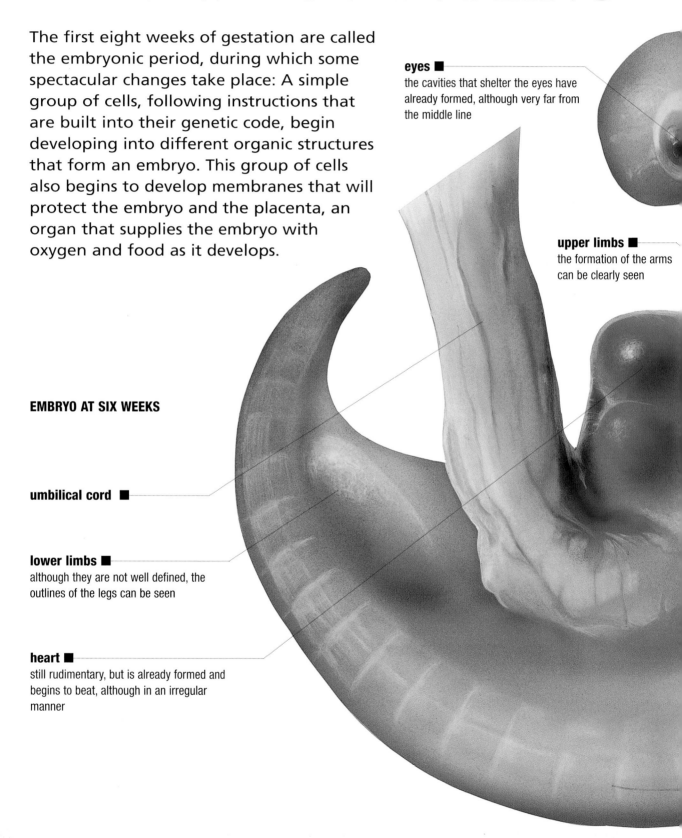

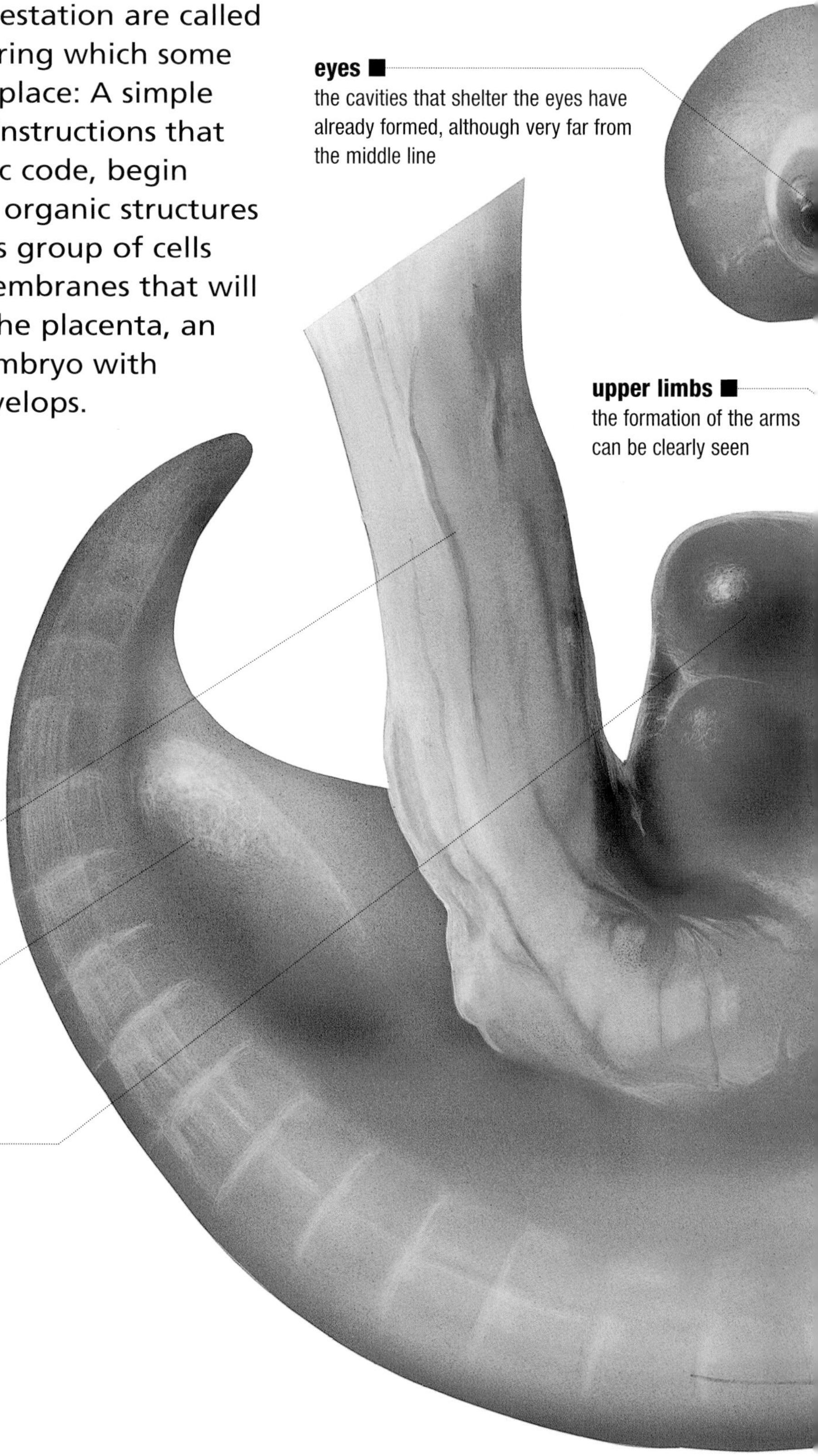
eyes
the cavities that shelter the eyes have already formed, although very far from the middle line
upper limbs
the formation of the arms can be clearly seen
umbilical cord
lower limbs
although they are not well defined, the outlines of the legs can be seen
heart
still rudimentary, but is already formed and begins to beat, although in an irregular manner

EMBRYO AT SIX WEEKS

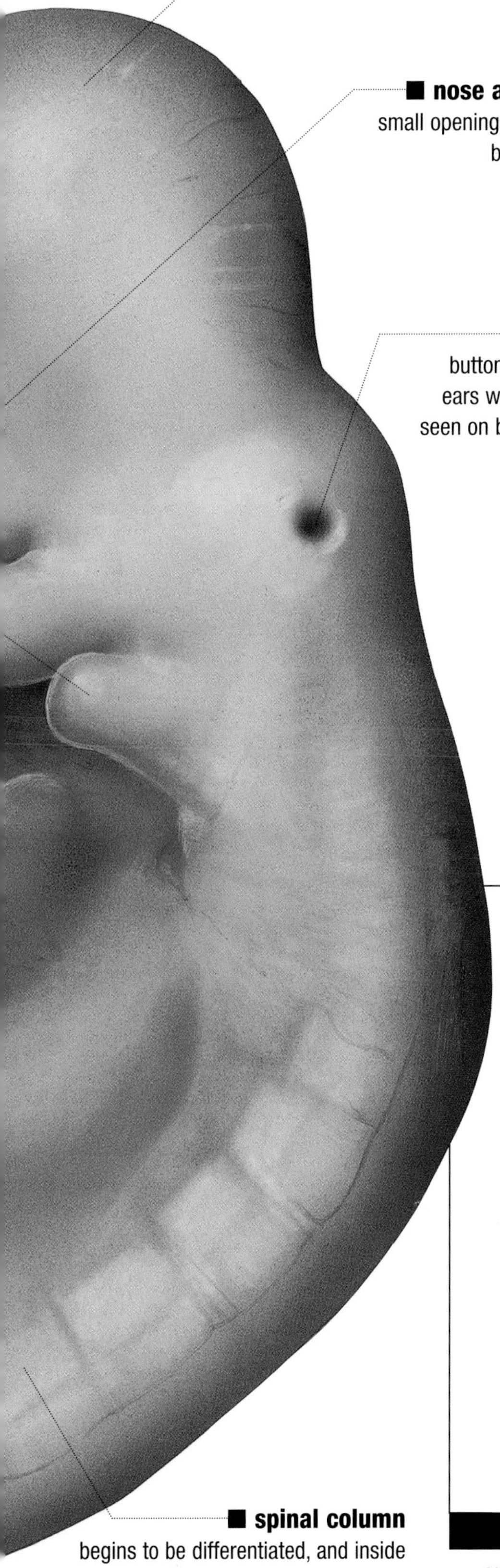

■ **head**
the prominence at the end of the embryo takes on the shape of a head, still very much folded in on the trunk

■ **nose and mouth**
small openings in the face begin to form

■ **ears**
buttons where the ears will be can be seen on both sides of the head

■ **spinal column**
begins to be differentiated, and inside it, the nerve tissue that will form the spinal medulla begins to develop

DEVELOPMENT OF THE PLACENTA

Inside the uterus, the embryo remains protected by membranes and surrounded by the warm fluid in which it floats. It is connected to the mother's body by the umbilical cord, through which pass the blood vessels that connect its tiny body with the placenta, an organ that acts as a bridge between mother and child. The placenta is a very special organ. It develops during pregnancy and is responsible for the exchange of substances between the mother's blood and that of the fetus, which are never in direct contact.

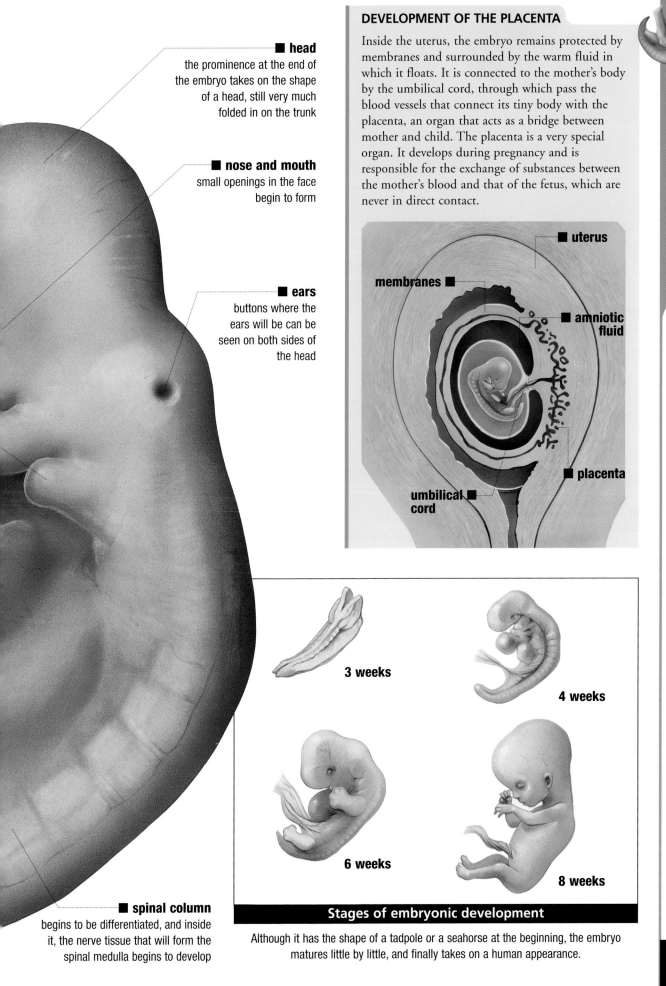

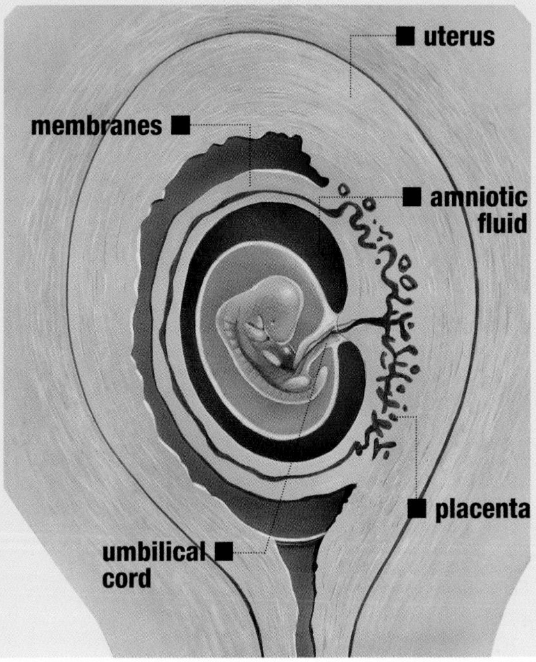

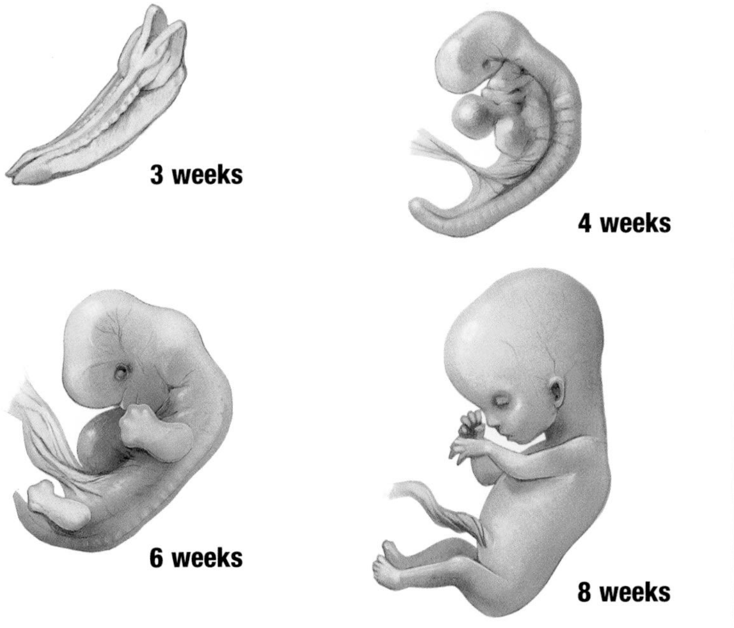

Stages of embryonic development

Although it has the shape of a tadpole or a seahorse at the beginning, the embryo matures little by little, and finally takes on a human appearance.

FROM THREE TO SIX MONTHS

Between the third and sixth month of gestation, the developing fetus has nearly all of its organs. From here on, the organs will continue to develop more fully until they have completely formed. The fetus increasingly takes on the appearance of a delicate baby, although it still needs the protection of its mother's womb before it can survive on its own in the outside world.

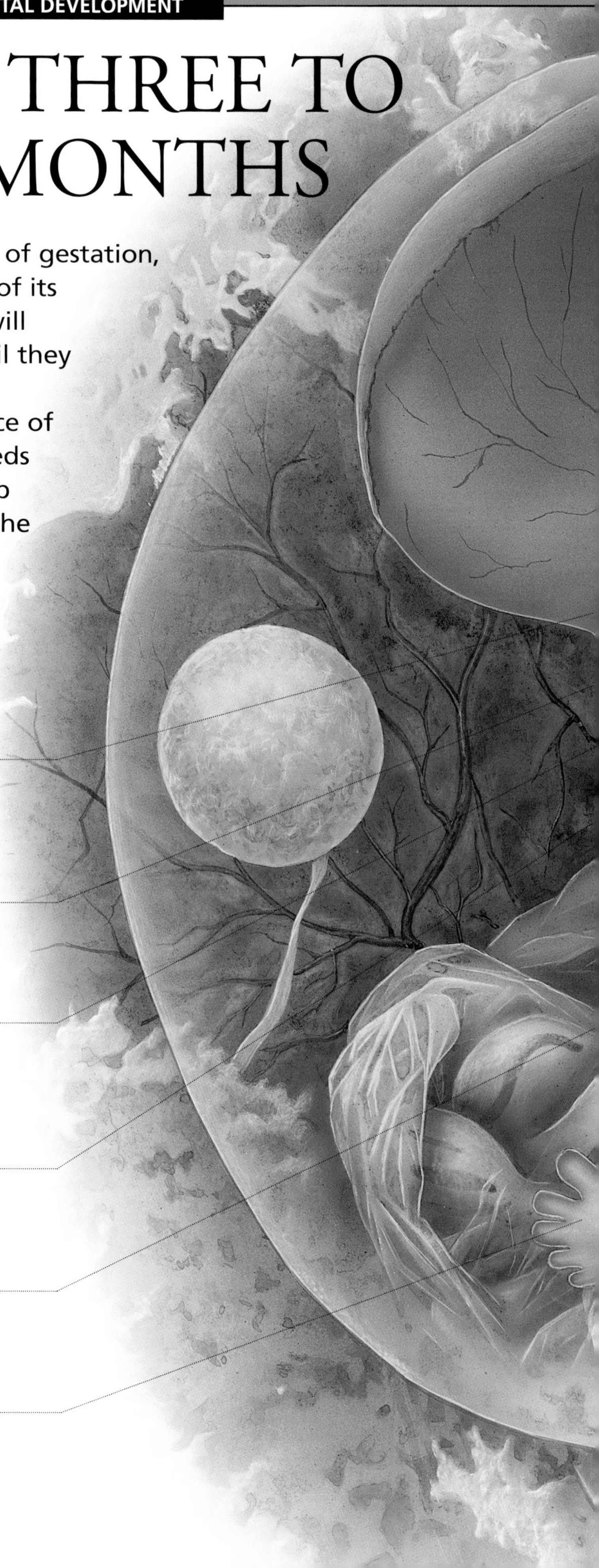

eyes ■
come to be located in front, although they are still widely separated; they are already covered with a fine layer of skin that will become the eyebrow

face ■
the face completes its definition, and its features take on a human appearance

mouth ■
the lips begin to define themselves, and the tongue and teeth are formed, although it will take a great deal of time before the teeth appear

hands ■
the fingers are already distinguished and will soon have nails and fingerprints

heart ■
beats without stopping and pushes blood through the circulatory vessels of the body

feet ■
the legs lengthen and the feet, with all their toes, are already formed

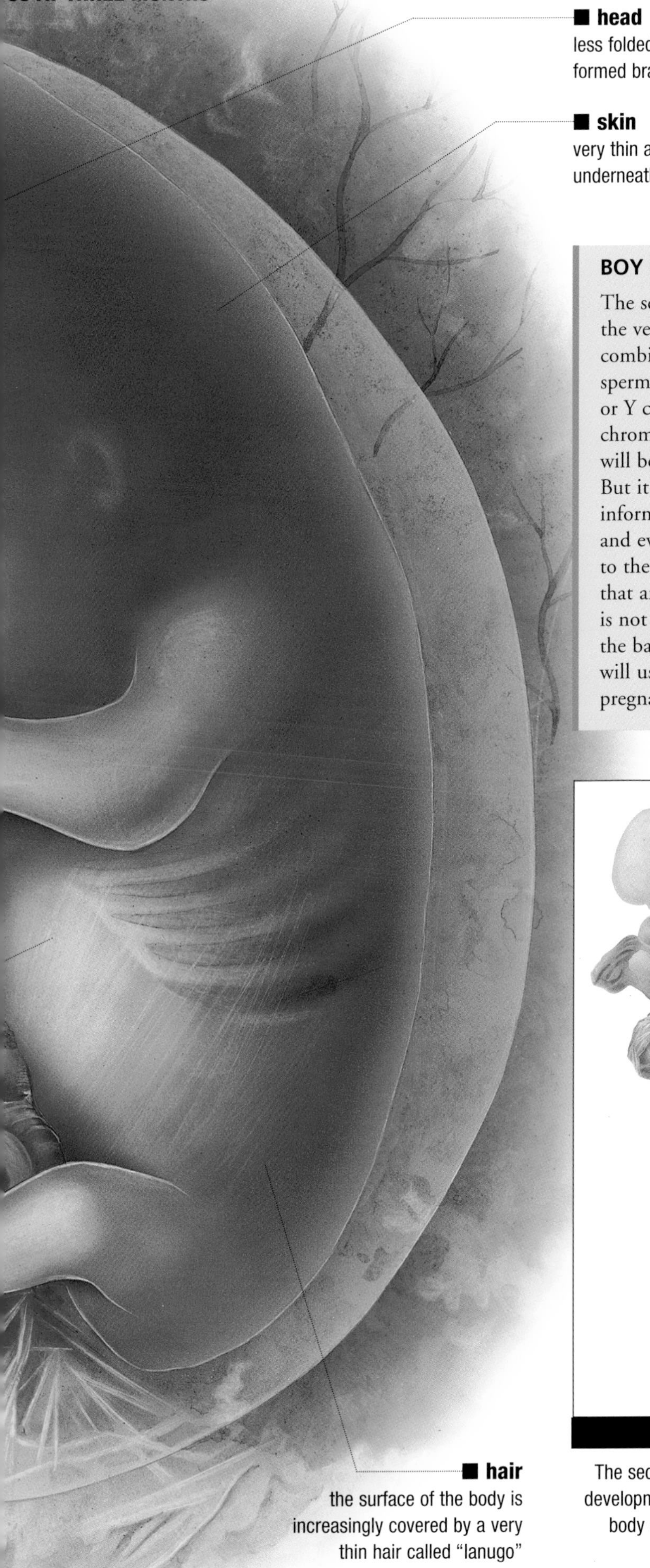
■ head
less folded over the trunk; almost has a fully formed brain
■ skin
very thin and delicate; allows the blood vessels underneath to show through
■ hair
the surface of the body is increasingly covered by a very thin hair called "lanugo"

BOY OR GIRL?

The sex of the new human being is determined at the very moment of fertilization, according to the combination of sex chromosomes brought by the sperm cell coming from the father (X chromosome or Y chromosome) and the mother's egg cell (X chromosome). If the combination is XX, the baby will be a girl; if it is XY, the baby will be a boy. But it still takes a few weeks before genetic information is transformed into bodily changes, and even longer for the genital organs to develop to the point that they can be identified in the tests that are conducted to monitor a pregnancy. But it is not necessary to wait until birth to know whether the baby is a boy or a girl: The gender of the baby will usually be apparent by the midpoint of a pregnancy.

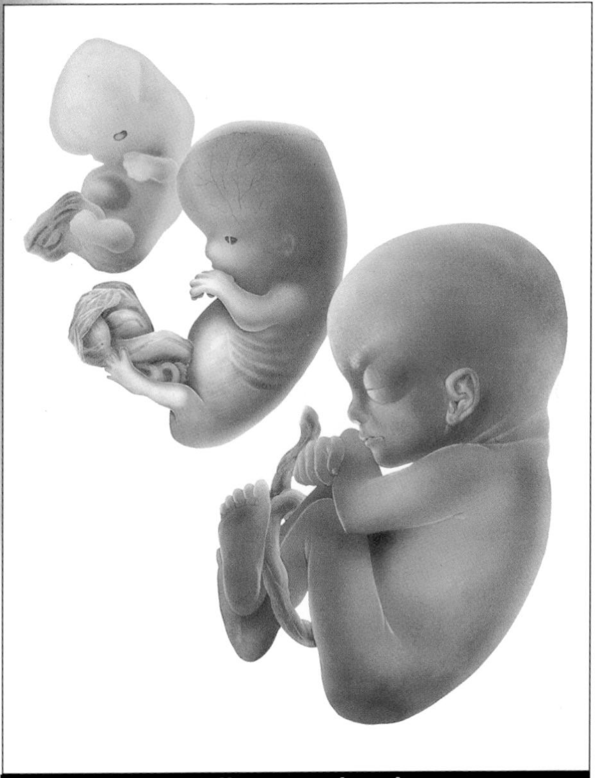

Full steam ahead

The second trimester of gestation is a crucial stage in fetal development, because the body grows at a rapid rate and the body structures take on definite human characteristics.

FROM SIX TO NINE MONTHS

The period from the sixth month of gestation until birth, the third trimester of pregnancy, represents a period of maturation. The body organs of the fetus have already fully developed, and need only to begin working at full capacity to ensure survival in the outside world.

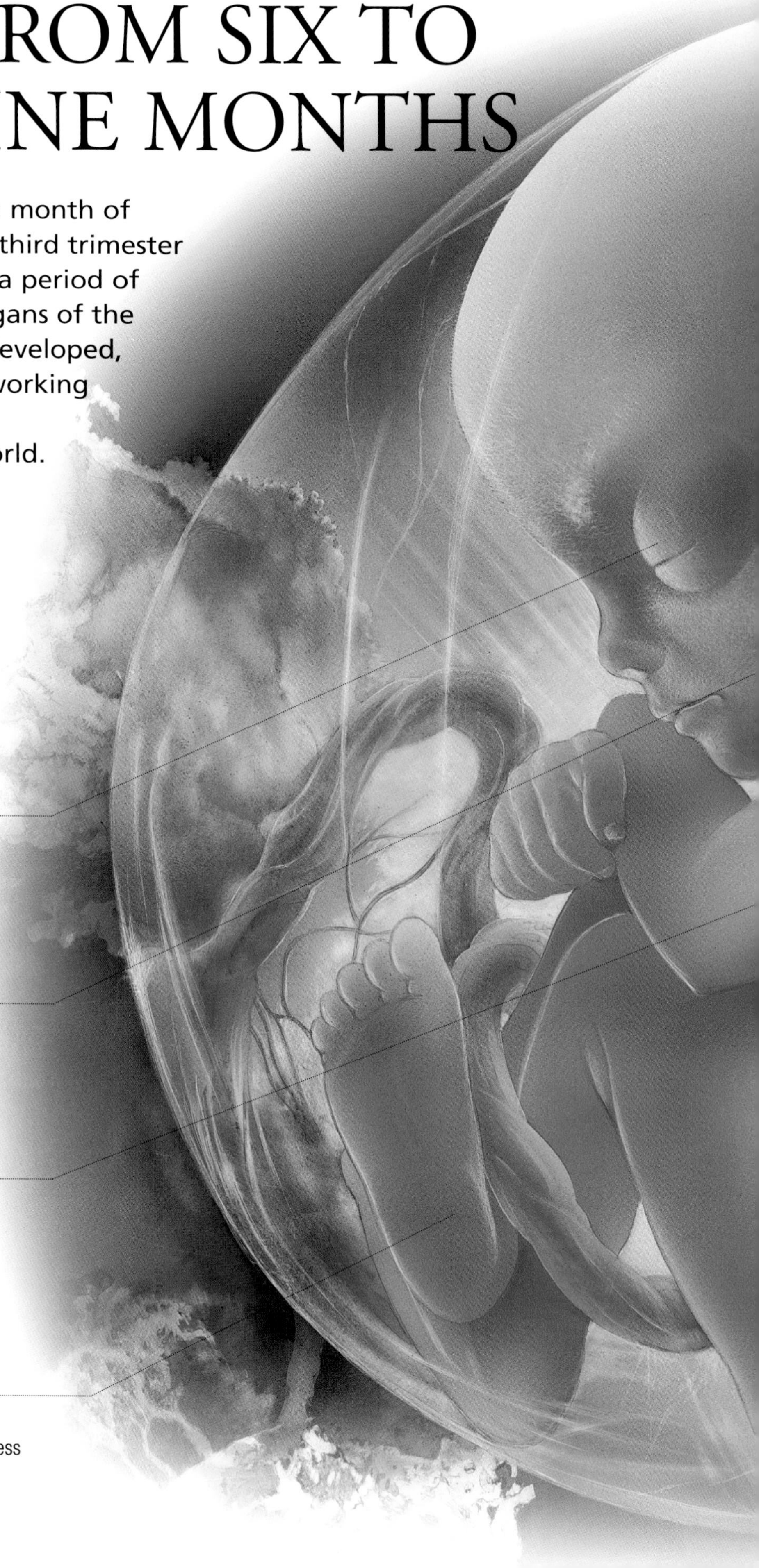

eyes ■
located in their final position, are covered by completely formed eyelids

face ■
the features are sharper and the profile of the nose better defined

hair ■
lanugo is replaced by vellus hair, and the brows and hair of the head begin to grow

position ■
the future baby adopts the typical "fetal position," with its limbs folded to take up less space

SIX-MONTH-OLD FETUS

■ head
of reasonable proportion compared with the body, and attached to the body with a more apparent neck

HOW WE GROW!

Head. At two months of fetal age, the head is half the size of the body; at birth, one-quarter; as an adult, one-eighth.
Weight. At two months of fetal age, the fetus weighs 0.28 ounces; at six months, the fetus weighs 100 times more (28 ounces); at birth, the baby weighs 400 times more (7 pounds), and as an adult, 8,700 times more (154 pounds).
Height. At two months of fetal age, the embryo is a little more than one inch long; at six months, the fetus measures 11 times more (just over a foot long); at birth, the baby measures 16 times more (20 inches); an adult is 56 times taller (5 feet, 6 inches).

■ ears
are already in their proper place, and earlobes are clearly seen

■ skin
the body surface is covered by a layer of fat that protects the fetus from continual contact with the liquid in which it is submerged

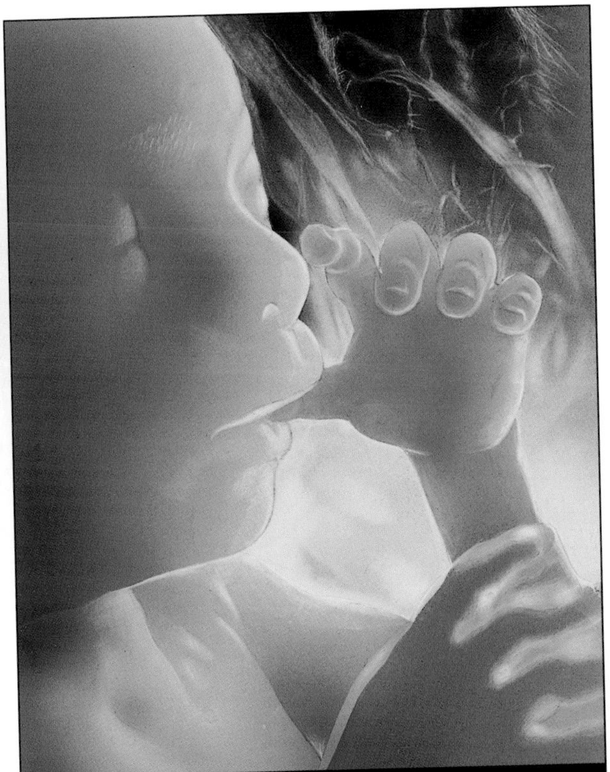

She's sucking her thumb!

At this stage, the nervous system of the fetus is already mature enough for the fetus to carry out some fairly complex actions: It moves around a lot—often described by the mother as "kicking"—and if we put our hand on her belly at this moment, we will certainly feel it. The fetus opens and closes its eyes, blinks, and reacts with brisk movements to very strong light or sound stimuli from outside. It even sucks its thumb, as if it were practicing to be able to feed itself after it is born.

WHEN MORE THAN ONE BABY BEGINS TO DEVELOP

Although gestation in humans usually results in the birth of a single child, it is also normal for two, or less frequently, three or even more babies, to be born from the same pregnancy. There are basically two types of twins or multiple births: those resulting from the splitting of one fertilized egg—identical twins (also called maternal twins); and those resulting from the fertilization and development of two different eggs—nonidentical twins (also called fraternal twins).

placenta ■
a single placenta, to which the umbilical cords of both fetuses attach, develops

chromosomal makeup ■
the fetuses have the same genetic information because they have the same chromosomal makeup, given that they come from the same sperm cell and the same egg cell

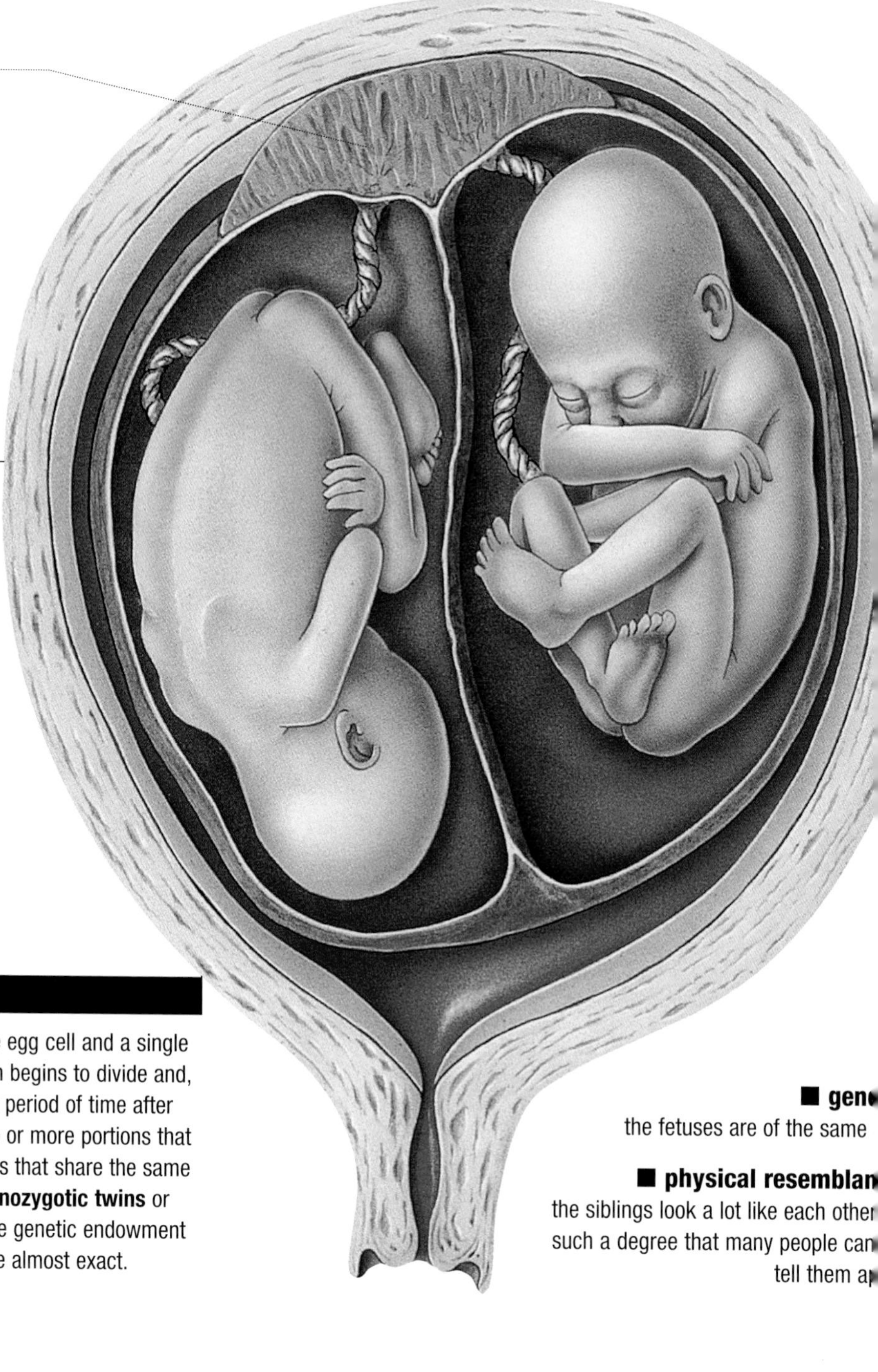

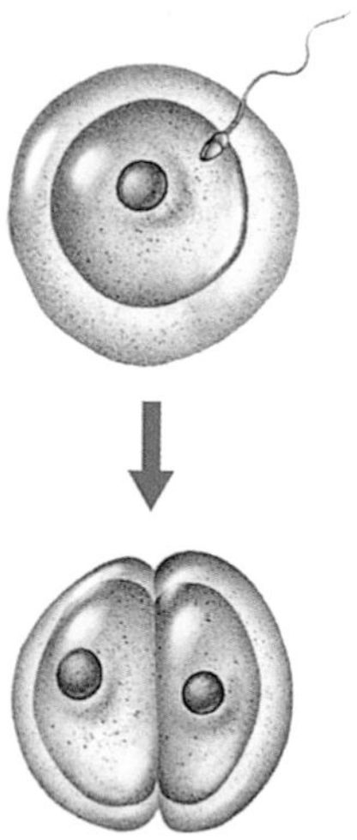

Identical twins

These twins came from the union of a single egg cell and a single sperm cell. The zygote created by fertilization begins to divide and, for unknown reasons, at the end of a short period of time after implanting in the uterus, it is broken into two or more portions that create multiple embryos. Two or more fetuses that share the same placenta develop. These are known as **monozygotic twins** or **identical twins** because they have the same genetic endowment and their physical resemblance will be almost exact.

■ **gen**
the fetuses are of the same

■ **physical resemblan**
the siblings look a lot like each other such a degree that many people can tell them a

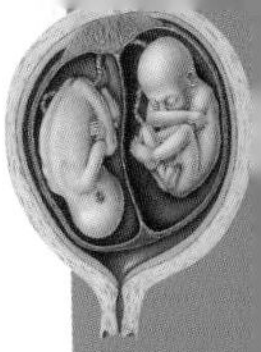

FREQUENCY

It is estimated that a multiple pregnancy occurs once for every 80 single-child pregnancies. Generally, only two twins result from multiple gestations, but triplets occur in one of every 8,000 pregnancies, quadruplets in one of every 750,000, and quintuplets in one of every 65,000,000. Pregnancies of six or seven fetuses are rare, although they do occur, and there are references to pregnancies of eight and even nine embryos.

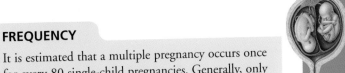

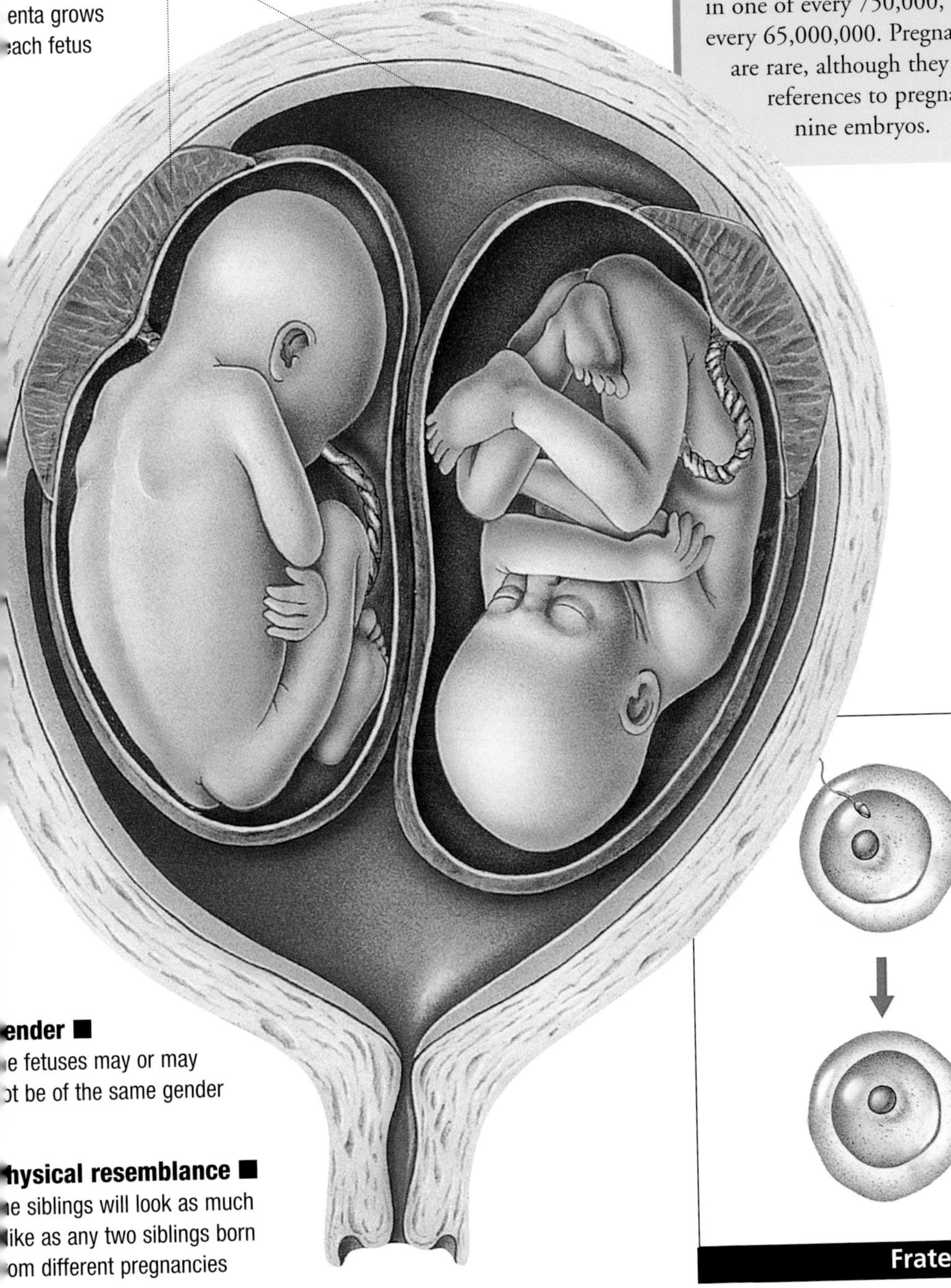
centas ■
parate
enta grows
ach fetus
■ chromosomal makeup
the fetuses will have different genetic information, unique for each one, because they come from different sperm cells and different egg cells
ender ■
e fetuses may or may
t be of the same gender
hysical resemblance ■
e siblings will look as much
ike as any two siblings born
om different pregnancies
ould

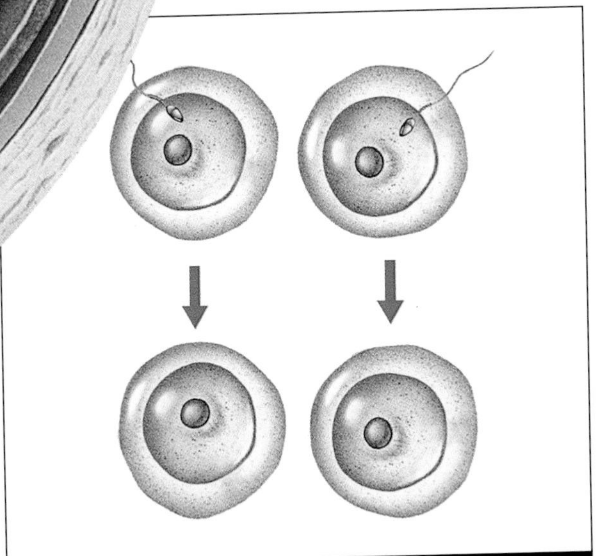

Fraternal twins

These twins result if the mother releases two or more egg cells during the same cycle, and these are fertilized at the same time by as many sperm cells. Thus, there are two or more zygotes that are transformed into several embryos. In such cases, two or more fetuses will be implanted independently into the uterus, and each will have its own placenta. They are known as **dizygotic twins** or **fraternal twins** because each has a different genetic endowment. They may or may not be of the same gender and their physical resemblance will not be identical.

FAMILIES OF TWINS

It is said that there are families with a predisposition to the birth of twins, and statistical data show that this is true. In some families, a greater tendency to have double ovulations is inherited from the mother; this considerably increases the probability of having twins. Thus, it is not strange that there might be several cases of twins among the women of the same family.

THE TIME OF DELIVERY

After nine months of pregnancy, the fetus has developed and matured enough to ensure the likelihood of survival in the outside world. The fully developed baby is ready to pass from the comfortable maternal womb to begin an independent life in the strange outside world. Nevertheless, the baby will still depend for a long time on the care of its parents.

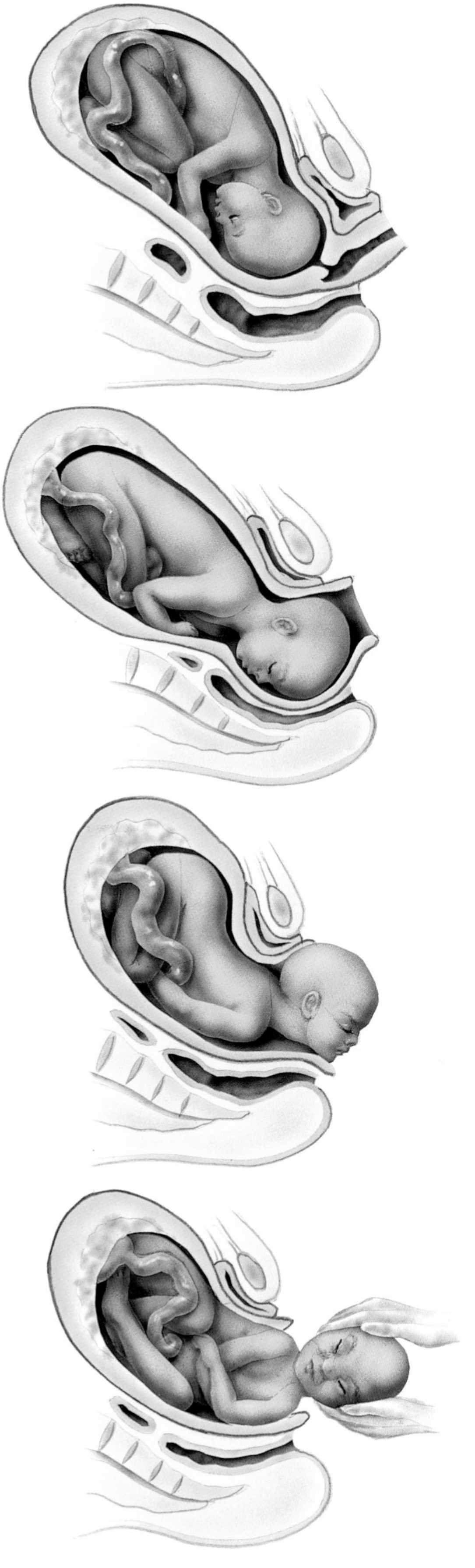

Delivery

THE CAESAREAN SECTION

There are occasions when, for various reasons, complications arise at the time of delivery that put the outcome of the pregnancy in danger. In these cases, sometimes the only way to save the baby or to speed up delivery is for the doctor or midwife to perform a surgical operation in which the baby is taken out through an incision in the abdomen of the mother. This operation, called a caesarean section, is very common today, and does not pose great harm to either the mother or the baby.

To leave the maternal womb and face the world, the baby must travel a path of four to five inches that separate it from the outside. This is made possible by powerful contractions of the muscular uterine walls of the mother. This process is known as delivery: The head comes out first, followed by the shoulders and the rest of the body.

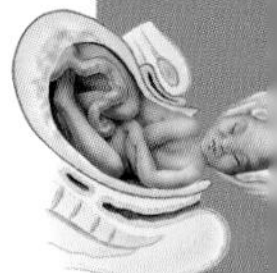

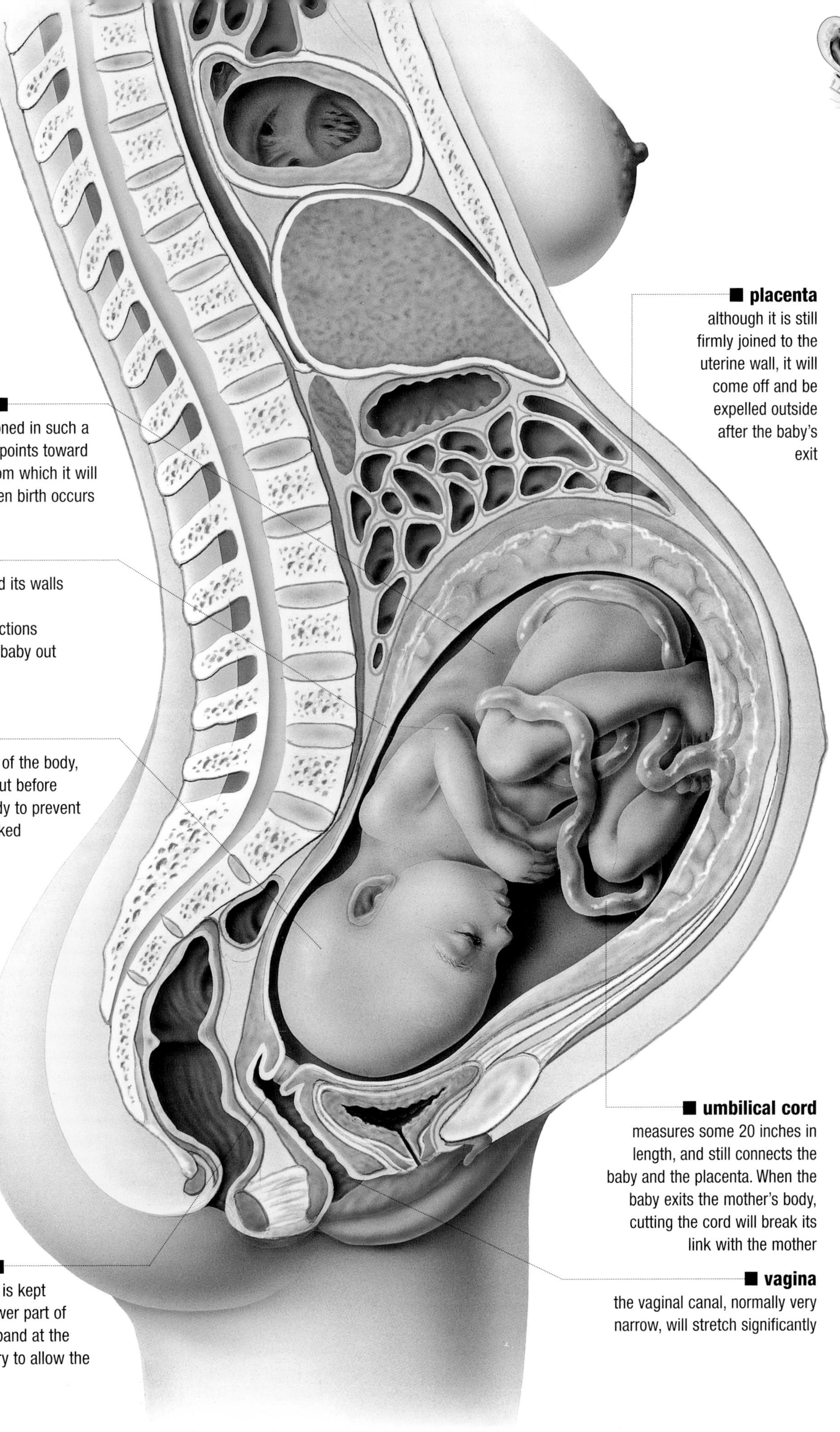

l position ■
baby is positioned in such a
that its head points toward
birth canal, from which it will
e out first when birth occurs

rus ■
ry dilated, and its walls
soon begin to
rience contractions
will push the baby out

d ■
e largest part of the body,
it will come out before
rest of the body to prevent
m being blocked

rine neck ■
uterine orifice is kept
ed, but the lower part of
uterus will expand at the
nent of delivery to allow the
to come out

■ placenta
although it is still firmly joined to the uterine wall, it will come off and be expelled outside after the baby's exit

■ umbilical cord
measures some 20 inches in length, and still connects the baby and the placenta. When the baby exits the mother's body, cutting the cord will break its link with the mother

■ vagina
the vaginal canal, normally very narrow, will stretch significantly

CHROMOSOMAL MAKEUP: THE INSTRUCTION MANUAL

The information needed for the development of a new human being, for both physical traits and bodily functions, is stored in genes contained in the chromosomal DNA contributed by the germ cells from the parents. The sperm cell and the egg cell both contribute 23 chromosomes, so that the fertilized egg has a complete set of 46 chromosomes.

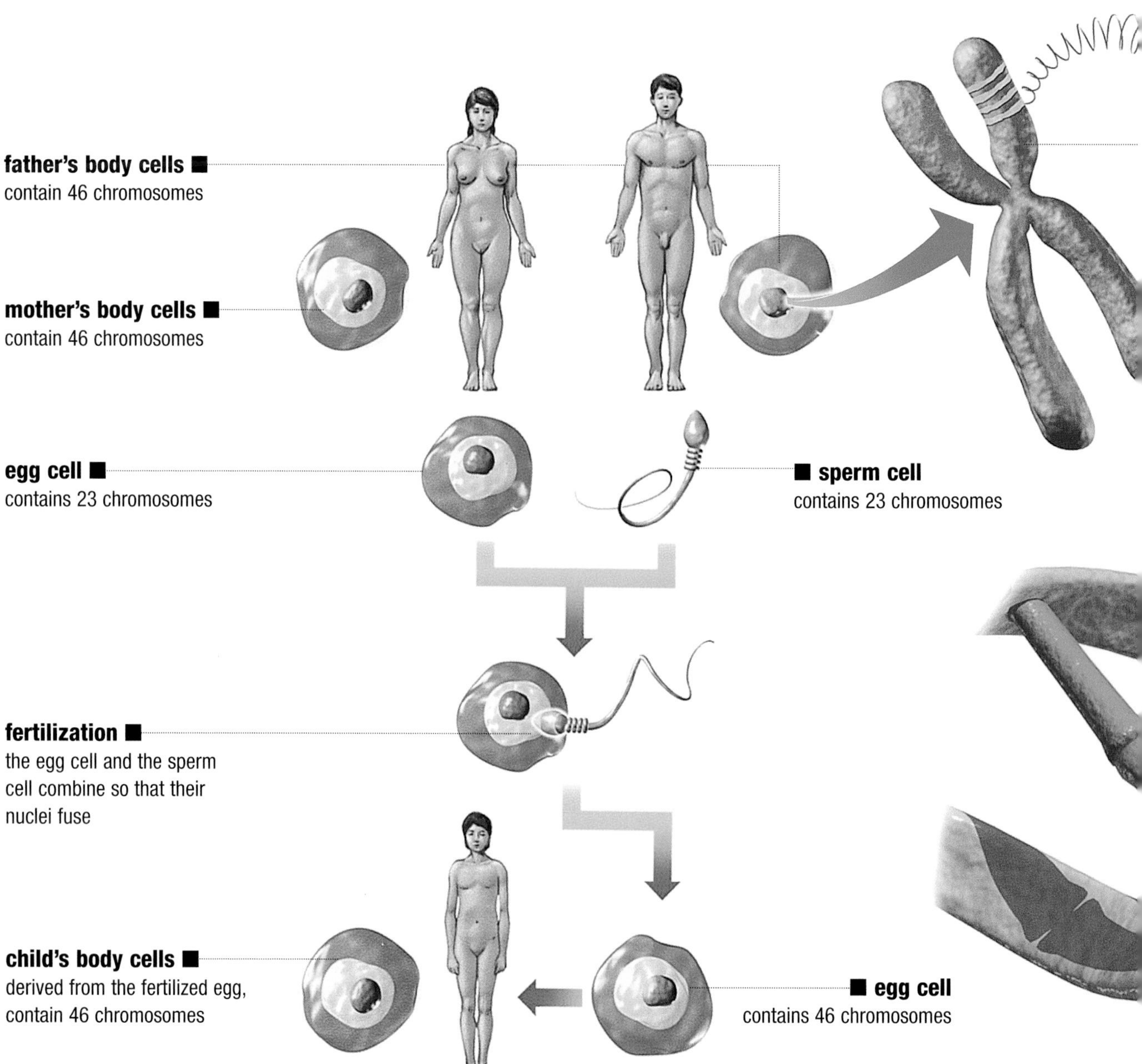

:HROMOSOMES AND SPECIES

:ach animal or plant species has a certain number of chromosomes, which ; transmitted from parents to offspring. There is no direct relationship ›etween the number of chromosomes a species has and its complexity. There are worms that have only two chromosomes and flies that have four. Tomatoes have 16, frogs 26, and bees 32. A human has 46 chromosomes, he same as an ash tree, while a gorilla has 48 and a dog has 78. There is a butterfly that has 446 chromosomes and a fern hat has no fewer than 1,360.

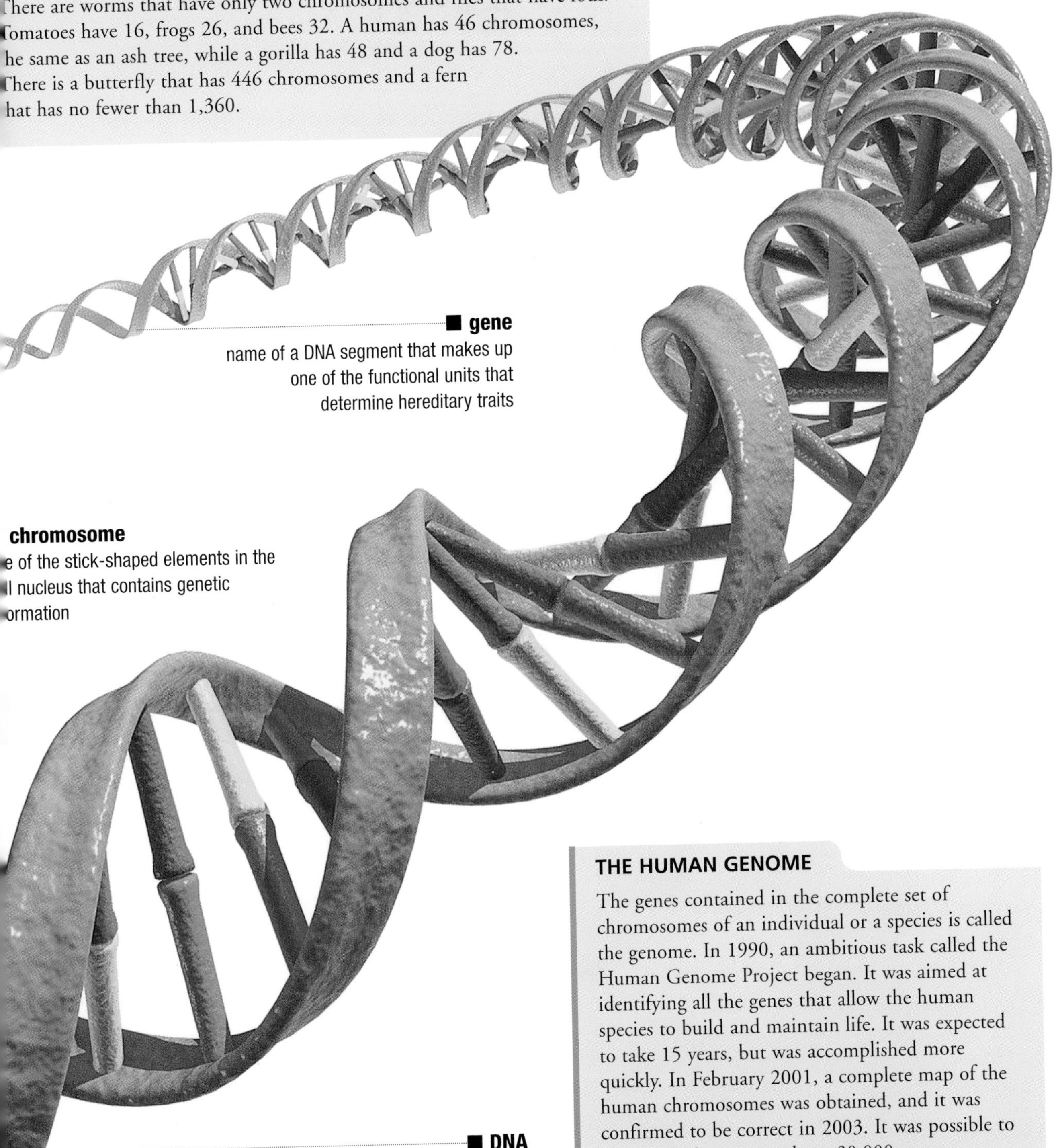

■ **gene**
name of a DNA segment that makes up one of the functional units that determine hereditary traits

chromosome
e of the stick-shaped elements in the l nucleus that contains genetic ormation

■ **DNA**
acronym for "deoxyribonucleic acid," the basic component of chromosomes and the substance that stores encoded genetic information

THE HUMAN GENOME

The genes contained in the complete set of chromosomes of an individual or a species is called the genome. In 1990, an ambitious task called the Human Genome Project began. It was aimed at identifying all the genes that allow the human species to build and maintain life. It was expected to take 15 years, but was accomplished more quickly. In February 2001, a complete map of the human chromosomes was obtained, and it was confirmed to be correct in 2003. It was possible to identify and sequence about 30,000 genes, although the exact function of many of these is not yet understood. These genes occupy only 3% of the chromosomal DNA. The remaining genes, which presumably exist to control the function of the ones identified, are still a mystery. We have made great advances in understanding our genome, but there is still much left to learn.

HEREDITY AND PEAS

The chromosomal makeup of each person is a chance combination of genes. We inherit half of our chromosomes from our fathers and the other half from our mothers, and we only pass on half of our chromosomes to our descendants. Although each person shares a certain portion of the chromosomal makeup of his or her parents, and there are usually similar traits among members of the same family, the chromosomal makeup of each person is unique and unrepeatable. Each one of us is special.

1. Gregor Mendel
Austrian botanist (1822–1894), grandson of a gardener and son of farmers, studied philosophy and entered an Augustinian monastery in 1843; he was in charge of the monastery garden for years.

2. Smooth and rough peas
Mendel's curiosity and reasoning ability led him to notice certain coincidences as he grew peas in the corner of his garden. He saw th when he planted rough peas, rough peas grew; when he planted sm peas, smooth peas grew; and when he crossed them, sometimes rou peas grew, and sometimes smooth peas grew.

3. Conclusions
Mendel crossed different types of peas, noted their combinations, and selected some traits over others—traits such as the shape of the pea, the height of the plant, or the color of the flowers. From his results, he drew a series of conclusions.

4. Mendel's laws
The work of the Augustinian monk is an example of scientific methodology that allowed Mendel to establish the general laws of inheritance, making it clear that the traits of children depend on "particles" that come from the parents, at a time when the existence of chromosomes was unknown.

5. Father of genetics
Mendel published his work in 1865, but it went unnoticed for several decades, until other investigators noticed the importance and accura of his conclusions and raised his work to the position of importance occupies today.

: GENETIC CODE

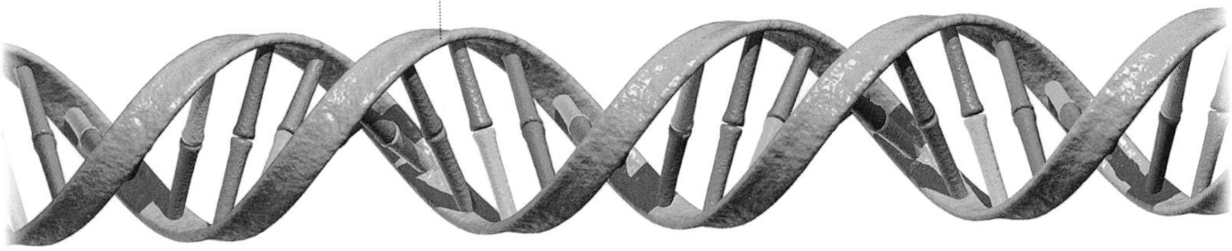

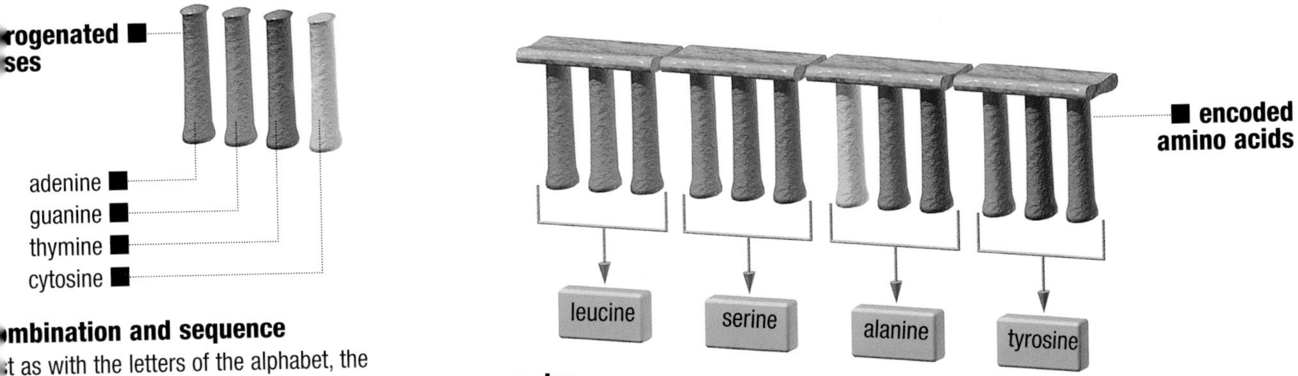

mbination and sequence
t as with the letters of the alphabet, the mbination of nitrogenated bases—their sequence the DNA chain—is very important because it rresponds to the instructions needed for making oteins, which are formed by a combination of ferent amino acids

codon
each sequence of three bases, called a codon, represents one of the different amino acids that make up the proteins of our bodies. The combination of these base "letters," which correspond to amino acids, give rise to the formation of "words," proteins, and indeed, to a whole "book," which is our body.

IE INHERITANCE OF EYE COLOR

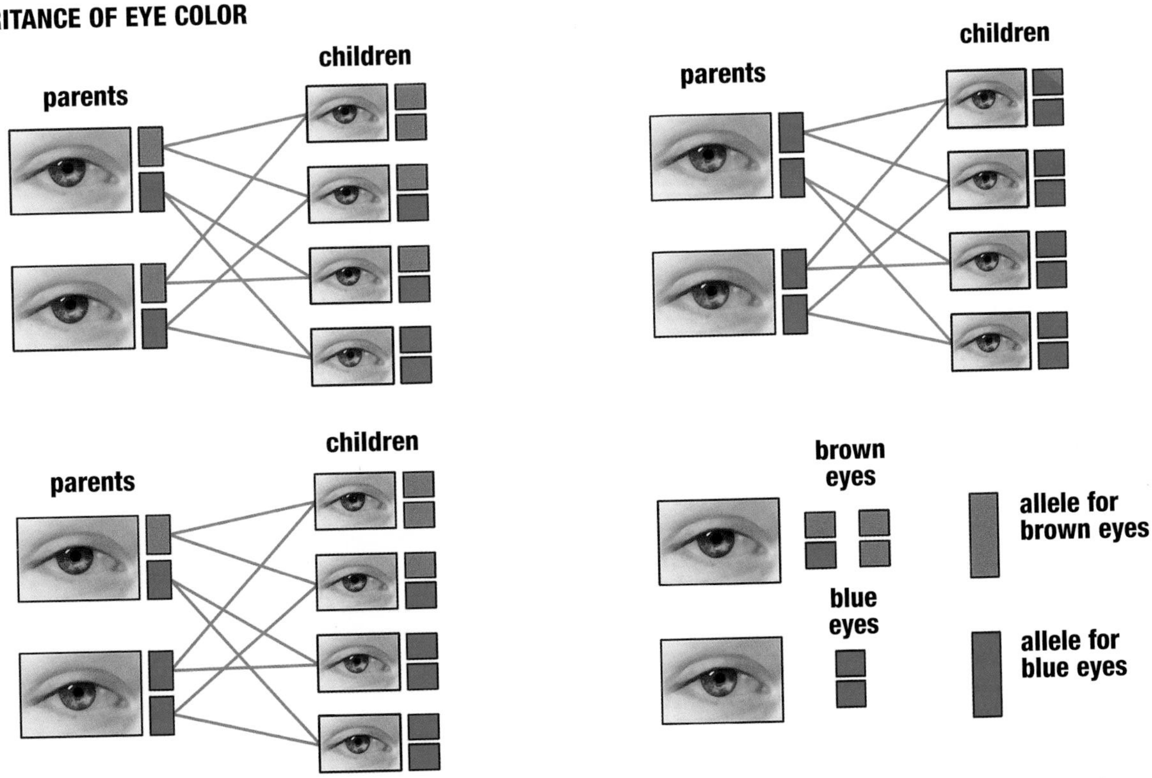

Many physical traits are encoded by a single gene. That gene exists, however, in different versions called **alleles**. Sometimes, in each chromosome pair—one inherited from the father and the other from the mother—there is a different allele, and it may happen that the information contained in one imposes itself over the other: The first is considered **dominant** and the second is **recessive**. This is what happens with the inheritance of eye color: The allele for brown eyes is dominant, and the one for blue eyes is recessive. For this reason, possible inheritance combinations are different. Although children whose mother and father both have brown eyes may have blue eyes (perhaps like one of their grandparents), the children whose parents both have blue eyes will only have blue eyes.

DID YOU KNOW?

	2 months	3 months	4 months	5 months	6 months	7 months	8 months	9 months
length	1.2–1.6 in	3.9 in	6.3 in	9.8 in	12.6 in	15.8 in	18.5 in	19.7 in
weight	.07–.11 oz	1.1 oz	5.3 oz	8.8–10.6 oz	1.3 lbs	2.6–3.3 lbs	4.4–5.5 lbs	6.6–7.7 lbs

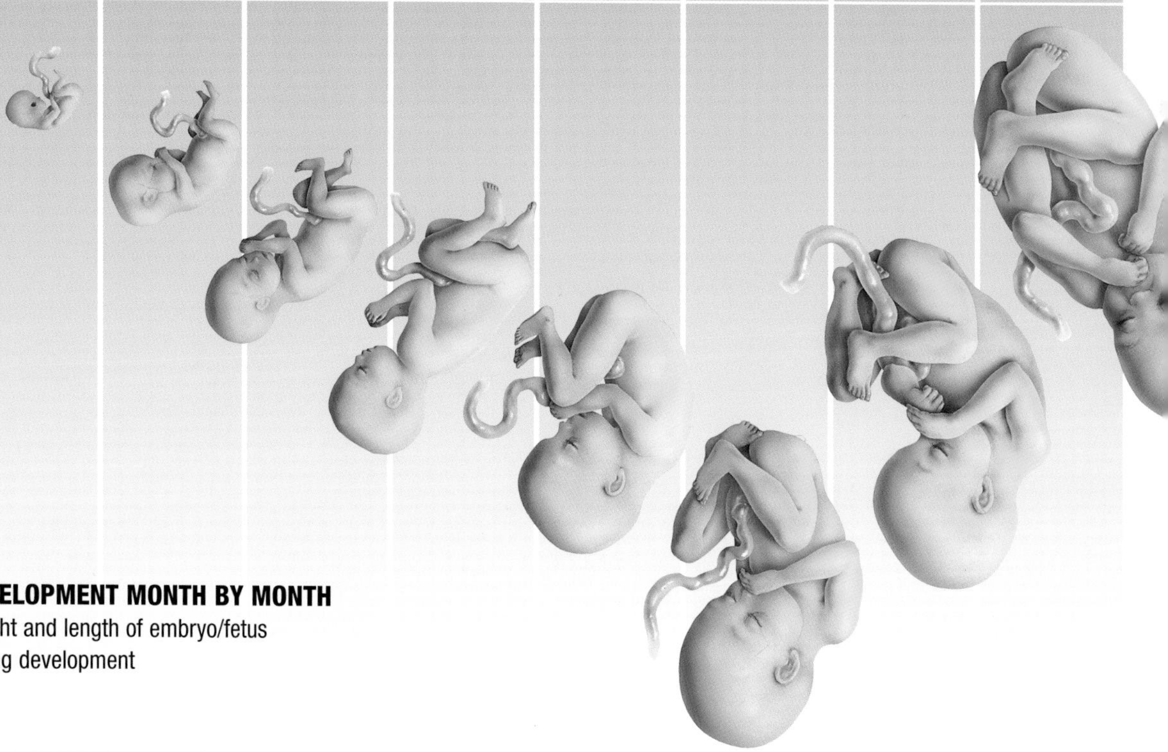

DEVELOPMENT MONTH BY MONTH
Weight and length of embryo/fetus during development

HOW LONG DOES PREGNANCY LAST?

In humans, pregnancy generally lasts 266 days from fertilization to birth. However, the duration of pregnancy is usually not calculated from the day of fertilization, but from the date on which the mother had her last menstrual period, since menstruation is interrupted during the entire period of pregnancy and the last period takes place two weeks before conception. According to this method, it is usually considered that pregnancy lasts 280 days—10 lunar months, or a little more than 9 calendar months. This is not just anecdotal information, since it can be used to make calculations so a mother can know the approximate delivery date so that she can be prepared for this major event.

Pregnancy does not necessarily follow a mathematical formula. A baby can be born completely developed four weeks before or up to two weeks after the usual nine-month gestation period. In some cases, pregnancy can be even longer or shorter. It is not uncommon for the baby to be delivered several weeks or even a few months early, in which case the baby is "premature": smaller and less developed than a baby that has completed the full term. A premature baby can have many developmental problems, but these can often be overcome with appropriate medical care.

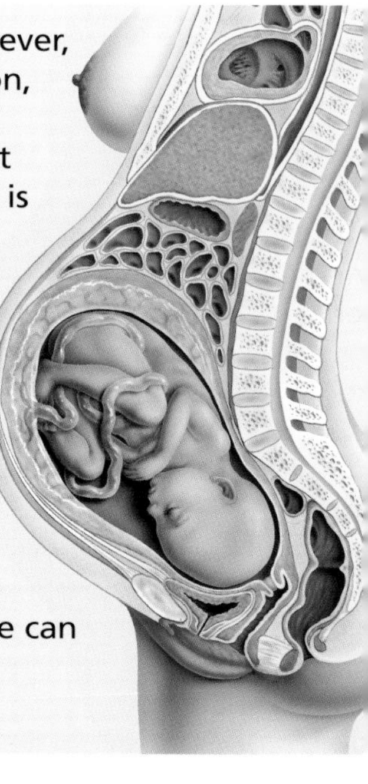

SONOGRAM

To check the progress of a pregnancy, there is a very useful diagnostic technique available today: the sonogram. This test, harmless for mother and child, is based on the use of ultrasounds, which are sound waves of a frequency so high that the human ear cannot hear them. Applied to the belly of the mother by means of a small emitter/receiver device, the ultrasound waves travel through the tissues and are reflected back in the form of an "echo." The echo's registration allows medical professionals to get sharp images of the fetus that can help them detect the baby's progress and health. With this technique, it is as if the doctor is able to examine the fetus while it is still inside its mother's uterus.

NO SMOKING

During pregnancy, the mother should avoid certain harmful habits that might seriously damage the development of the child, such as tobacco smoking, drinking alcoholic beverages, and taking drugs. For the duration of her pregnancy, she must be careful with the medicines she uses, because some medicines can be dangerous for the fetus. Before taking any medicine, she should consult a doctor.

INTRAUTERINE SURGERY

Advanced techniques available today allow doctors to identify problems and perform surgical procedures that would have been considered impossible just a few decades ago. An example of this is intrauterine surgery—operations performed on an unborn fetus while it is still inside the mother's womb. Instruments specially designed for this purpose, including the use of tiny video cameras that are introduced into the womb so that doctors can see the fetus on a monitor, make it possible to perform operations to correct problems that put the life of the baby in danger.

TALKING TO THE BABY

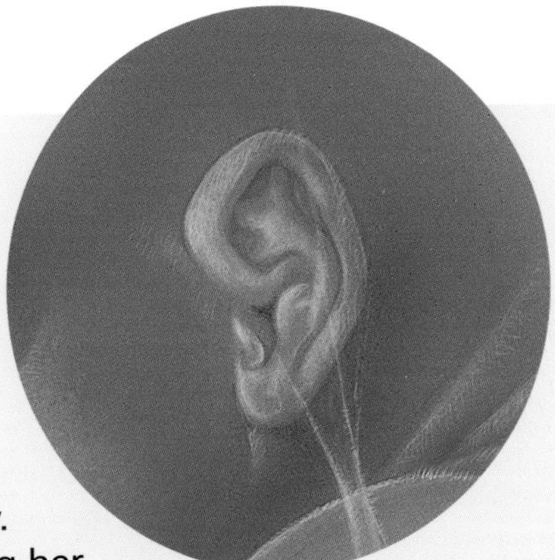

The fetus begins to develop the sense of hearing at around five months of pregnancy, or perhaps a little earlier. Of course, the sense will take time to reach full development, but it will soon allow the fetus to connect with the outside world. The fetus will begin to react to very loud sounds. Of course, in the shelter of its mother's womb, the sounds it hears are limited to those produced in its mother's body, especially the beating of her heart, which is a constant and powerful presence, and the mother's voice, which is transmitted by means of the amniotic fluid that surrounds the baby. Although the mother doesn't think about it, because she is directing her words to others, the baby hears her talk, sing, ask, order, complain, and laugh. And it is likely that the baby also hears when the father or siblings talk very close to the mother's belly.

INDEX